BURLEIGH DODDS SCIENCE: INSTANT INSIGHTS

NUMBER 38

Improving crop disease management

W0038640

burleigh dodds
SCIENCE PUBLISHING

Published by Burleigh Dodds Science Publishing Limited
82 High Street, Sawston, Cambridge CB22 3HJ, UK
www.bdspublishing.com

Burleigh Dodds Science Publishing, 1518 Walnut Street, Suite 900, Philadelphia, PA 19102-3406, USA

First published 2021 by Burleigh Dodds Science Publishing Limited

© Burleigh Dodds Science Publishing, 2021, except the following: The contribution of Dr Kelly
Turkington in Chapter 1 is © Her Majesty the Queen in Right of Canada. All rights reserved.

This book contains information obtained from authentic and highly regarded sources. Reprinted
material is quoted with permission and sources are indicated. Reasonable efforts have been made to
publish reliable data and information but the authors and the publisher cannot assume responsibility
for the validity of all materials. Neither the authors nor the publisher, nor anyone else associated with
this publication shall be liable for any loss, damage or liability directly or indirectly caused or alleged
to be caused by this book.

No part of this publication may be reproduced, stored in a retrieval system or transmitted in any
form or by any means electronic, mechanical, photocopying, recording or otherwise without the prior
written permission of the publisher.

The consent of Burleigh Dodds Science Publishing Limited does not extend to copying for general
distribution, for promotion, for creating new works, or for resale. Specific permission must be
obtained in writing from Burleigh Dodds Science Publishing Limited for such copying.

Permissions may be sought directly from Burleigh Dodds Science Publishing at the above address.
Alternatively, please email: info@bdspublishing.com or telephone (+44) (0) 1223 839365.

Trademark notice: Product or corporate names may be trademarks or registered trademarks and are
used only for identification and explanation, without intent to infringe.

Notice
No responsibility is assumed by the publisher for any injury and/or damage to persons or property as
a matter of product liability, negligence or otherwise, or from any use or operation of any methods,
products, instructions or ideas contained in the material herein.

British Library Cataloguing in Publication Data
A catalogue record for this book is available from the British Library

ISBN 978-1-80146-169-6 (Print)
ISBN 978-1-80146-170-2 (ePub)

DOI 10.19103/9781801461702

Typeset by Deanta Global Publishing Services, Dublin, Ireland

Contents

© Burleigh Dodds Science Publishing Limited, 2021. All rights reserved.

4   Diseases affecting grain legumes and their management   75
     Keith Thomas, University of Sunderland, UK

© Burleigh Dodds Science Publishing Limited, 2021. All rights reserved.

Series list

© Burleigh Dodds Science Publishing Limited, 2021. All rights reserved.

© Burleigh Dodds Science Publishing Limited, 2021. All rights reserved.

Chapter 1

The role of crop rotation, intercropping and tillage practices for foliar disease management of wheat and barley

T. K. Turkington, Agriculture and Agri-Food Canada, Canada; K. Xi, Alberta Agriculture and Forestry, Canada; and H. R. Kutcher, University of Saskatchewan, Canada

1 Introduction

There have been a range of studies assessing the factors which predispose modern agriculture to biotic stresses such as plant diseases (Apple, 1977; Barnes, 1964; Cowling, 1978; Harlan, 1972; Paddock, 1967; Zadoks and Schein, 1979; Zhan et al., 2015). Boudreau and Mundt (1997) have characterized modern crop production as highly disrupted and simplified agro-ecosystems in which the stabilizing mechanisms that function in natural ecosystems have been significantly weakened. This can be seen, for example, in the use of uniform cultivars with reduced genetic diversity (Cowling, 1978; Harlan, 1972; Zadoks and Schein, 1979; Zhan et al., 2015). The narrow genetic basis for disease resistance in modern cultivars increases the potential for pathogens to overcome this resistance (Harlan, 1972). Modern agriculture results in greater plant uniformity, higher planting density and reduced rotations (which increased spatial and temporal uniformity) (Apple, 1977; Harlan, 1972; Wolfe, 1978; Zadoks and Schein, 1979; Zhan et al., 2015). This results in large blocks of genetically and morphologically uniform plants grown close together, resulting in dense crop canopies with stable microenvironments, which can allow pathogens to adapt quickly and attack a large number of plants at a similar state of development (Zadoks and Schein, 1979).

To counteract the predisposition of modern agro-ecosystems to plant diseases it has been suggested that cropping systems try to emulate some of the stabilizing

http://dx.doi.org/10.19103/AS.2018.0039.21
Published by Burleigh Dodds Science Publishing Limited, 2019.

characteristics of natural ecosystems (Apple, 1977; Zadoks and Schein, 1979). Zadoks and Schein (1979) have described a natural ecosystem as one in which 'disease is always present, but the host's genetic makeup, the population's genetic diversity and the dispersion of genotypes combine to prevent extreme disease growth rates'. Diversity is an integral component of a natural ecosystem that can be used to provide greater stability in modern agro-ecosystems with regard to disease management. In this chapter, diversity in time and space will be discussed in relation to managing cereal leaf diseases, primarily in wheat and barley.

2 Increasing temporal diversity: crop rotation

Karlen et al. (1994) have described the shift away from the use of legumes and other crops in rotations after the Second World War, and the increasing reliance on synthetic fertilizers and pesticides to lessen the need for management via 'extended rotations'. Farmers increasingly concentrated on growing a few crop species in a 'crop monoculture', sometimes referred to as a fixed cropping system or rotation (Finckh and Wolfe, 1997; Black et al., 1974; Tanaka et al., 2002). Because they are highly simplified, with a small number of crop components, these systems are more vulnerable to abiotic (weather, soil conditions etc.) and biotic (diseases, insects etc.) stresses (Finckh and Wolfe, 1997; Tanaka et al., 2002; Wolfe, 2002).

Crop rotation can be one of the most effective and sustainable methods of managing a number of plant diseases. However, it must be stressed that important factors such as relative commodity prices, net returns, ease of production/marketing, availability of storage facilities and on-farm feed/forage requirements are also very important considerations for farmers when planning crop rotations of sufficient length for disease management (Bailey et al., 2000; Cook, 2000; Cook and Veseth, 1991; Harker et al., 2015; Johnston et al., 2005; Karlen et al., 1994; Kutcher et al., 2011, 2013; Smith et al., 2013, 2015; Tanaka et al., 2002).

Crop rotation can be thought of as a biological method of disease management (Cook and Veseth, 1991). Rotation to a non-host(s) for a sufficient period allows enough time for decomposition of infected crop residues, and/or a reduction in the viability of pathogen survival structures and the pathogen's ability to produce inoculum, thus eliminating a potential source of disease. As Cook and Veseth (1991) indicated in their book *Wheat Health Management* '. . . rotation allows time for natural enemies to destroy the pathogens of one crop while one or preferably two unrelated crops are grown'. They also indicated that rotation acts like a natural type of 'soil fumigation' where the collective activity of 'antibiotic, predatory, and competitive organisms' help to eliminate plant pathogens from soil and infected crop residues.

The success of rotations as a form of biological disease control is dependent on a number of factors including time, the environment, the nature of the pathogen being managed and the characteristics of the host crop. For rotation to be an effective disease management strategy, farmers need to keep in mind that certain requirements or conditions must be met. Firstly, the majority of the disease inoculum must be from the field itself with limited movement of plant pathogen inoculum from adjacent fields. Consequently, crop rotation will tend to be most effective for those diseases that are soil- or residue-borne. Crop rotation will not be effective for diseases where seed-borne pathogens are an important source of disease, such as the cereal smuts. Furthermore,

Published by Burleigh Dodds Science Publishing Limited, 2019.

rotation will not be effective for cereal rusts as their causal agents are mobile and readily move from one field to another or over long distances. For example, cereal rusts, especially for wheat and barley, will overwinter on cereals and grasses in the southern United States and northern Mexico, while stripe rust can also overwinter in the Pacific Northwest (PNW) and California (Mathre, 1997; Menzies and Gilbert, 2003; Tekauz, 2003; Wiese, 1987). Urediniospores are blown northward by wind currents, affecting successive northerly winter and spring cereal crops (Aylor, 1990, 2003; Brown and Hovmøller, 2002; Chen, 2005; Eversmeyer et al., 1984; Eversmeyer and Kramer, 2000; Menzies and Gilbert, 2003; Nagarajan and Singh, 1990). However, viable overwintering inoculum in winter wheat is considered to play a role in causing local epidemics of wheat stripe rust (*Puccinia striiformis* Westend.) in central Alberta (Kumar et al., 2013). Thus, rotation away from wheat and barley may help manage stripe rust in these types of regions. Similarly, in winter-grown short season crops, as in Australia, the elimination of the 'Green Bridge', susceptible hosts growing during the summer, can substantially reduce disease levels.

A second major consideration regarding the use of a diverse crop rotation to manage disease is that the host range of the pathogen must be narrow. Crop rotation amongst different small grain cereals such as wheat, barley and oat as well as corn is an effective strategy for management of leaf spot diseases including tan spot, the septorias, scald and the net blotches. However, rotations amongst these crops is not an effective strategy for a disease like *Fusarium* head blight, caused by *Fusarium graminearum* Schwabe *sensu lato*, as this pathogen has an extensive host range that includes all these small grain cereals and corn (Mathre, 1997; Wiese, 1987; Sutton, 1982; Parry et al., 1995; Stack, 2000; Farr et al., 1989; Ginns, 1986). The final condition for rotation to be effective is that pathogens are unable to survive for long periods in the absence of a suitable host. Thus, there should be a rapid reduction in pathogen viability and its capacity to produce infective structures, that is spores, within a relatively short period of time. Rotation will not be as effective for those pathogens that have the capacity to produce long-lived resting structures or where they can survive in infected plant tissues that resist decay. For example, survival of *F. graminearum* in crop residue is highest in plant tissues that are resistant to decay, especially the node tissues of small grain cereals (Sutton, 1982; Burgess and Griffin, 1968).

2.1 Limiting cereal leaf diseases using crop rotation

Crop rotation can be an effective tool to lower the risk of cereal leaf diseases in wheat and barley (Bailey et al., 2000; Johnston et al., 2005; Krupinsky et al., 2006; Kutcher et al., 2011; Menzies and Gilbert, 2003; Tekauz, 2003; Turkington et al., 2006, 2012; Wiese, 1987). Some studies have shown that rotation is not always effective. As an example, Bailey et al. (2000) and Johnston et al. (2005) found that planting barley on non-host residues did not necessarily decrease the risk of net blotch of barley. In contrast, Turkington et al. (2006) found that in commercial barley fields in Alberta, from 1995 to 1997, lower levels of scald (*Rhynchosporium secalis* (Oudem.) J. J. Davis) and net blotch (*Pyrenophora teres* Drechs.) tended to occur when barley was planted in fields previously cropped to a non-host.

Krupinsky et al. (2004) found lower barley leaf spot levels, primarily net blotch, when barley was grown on non-host residue, especially wheat, canola and field peas. In a large crop sequencing study, Krupinsky et al. (2006) found that barley productivity was reduced when grown on barley residues versus non-host residues. More recently, Kutcher et al. (2011) found increased barley leaf spot severity, primarily net blotch, when barley was grown on barley residue versus non-host residues.

Turkington et al. (2012) have also found the largest impact on the barley leaf spot complex resulted from crop rotation versus fungicide application or nitrogen fertilizer rate. The largest reductions in disease and increases in grain yield and kernel size resulted from planting barley on either field pea or canola residue versus barley on barley residue. In general, for scald a rotation of 1–2 years away from a susceptible host may be sufficient to allow for the destruction of most infected plant tissue, since the scald pathogen primarily infects leaves and leaf sheaths, which decompose fairly rapidly (Mathre, 1997; Mayfield and Clare, 1984a,b). However, a longer rotation will be needed for net blotch, since the fungal pathogen that causes this disease also infects barley stems, which are more resistant to decay and will persist, especially the node tissue, for a longer period of time (Duczek et al., 1999; Mathre, 1997).

Crop rotation is also important in the management of wheat leaf spot diseases. Bailey et al. (1992) found that leaf spot severity was higher in wheat following spring or winter wheat compared with field peas or a summer fallow period. Bockus and Claassen (1992) found reduced tan spot (*Pyrenophora tritici-repentis* (Died.) Drechsler) levels with wheat grown on sorghum versus wheat residues, while wheat yields increased with rotation to sorghum, but only when adequate moisture levels occurred following the sorghum crop. In the Saskatchewan Parkland, Pedersen and Hughes (1992) found that when environmental conditions were favourable for disease development *Septoria nodorum* blotch (*Parastagonospora nodorum* (Berk.) Quaedvlieg, Verkley, and Crous) severity was reduced in wheat by a rotation with 2 years between wheat crops, but not for rotations with only 1 year between wheat crops. However, when environmental conditions were less conducive to disease, a rotation with 1 year out of wheat reduced *Septoria nodorum* blotch severity to the same extent as a rotation with 2 years between wheat crops.

In research trials and in commercial wheat fields in southwestern Saskatchewan, Fernandez et al. (1998) reported that leaf spot severity was not always lower in wheat planted after a single non-host compared with continuous wheat. They concluded that at least 2 years out of wheat were required for decomposition of infected residues to reduce wheat leaf spot levels. In a subsequent survey of commercial wheat fields, Fernandez et al. (2009) found a limited effect of rotation on wheat leaf spot severity, while in another study Fernandez et al. (2016) found reduced leaf spot severity when wheat followed a non-host or fallow.

Bailey et al. (2001) reported reduced leaf disease when wheat was grown on summer fallow or field pea residue versus wheat residue, and that wheat yields were increased when wheat was not grown on wheat residues. Krupinsky et al. (2004) found lower wheat leaf spot levels (tan spot and *Septoria nodorum* blotch), when wheat was grown on non-host residue, especially canola, barley and flax, but yields tended to be similar to wheat on wheat residue. In a large crop sequencing study, Krupinsky et al. (2006) also found reduced wheat productivity when grown on wheat residues versus non-host residues. Johnston et al. (2005) found that increased leaf spot severity and reduced yields were observed when wheat was planted on wheat residues versus a non-host.

In a review of the impact of break crops on wheat, Angus et al. (2015) reported improved wheat yields following non-host break crops versus wheat on wheat based on a combined analysis of data from Australia, Europe and North America. They focused on the potential impact of break crops primarily on root diseases as well as the supply of nitrogen and moisture. In Latvia, Bankina et al. (2015) found that rotation of wheat with non-host crops reduced tan spot, but was less effective for *Septoria tritici* blotch, except when there was more than 1 year between wheat crops. In subsequent research, Bankina et al. (2018) also found increased tan spot levels under continuous wheat versus rotations with non-hosts, especially under reduced tillage with a disc implement versus ploughing. Andert et al. (2016) investigated the impact of crop rotation on the intensity of fungicide use in Germany for a range of crops and found that continuous production of the same crop type for 3 years increased use intensity in winter wheat, winter barley and fall rye versus more diverse rotations. In Uruguay, Mazzilli et al. (2016) analysed the impact of the previous crop on wheat yields and found yield losses associated with planting wheat on wheat residue. However, yields increased when a disease-resistant cultivar was grown and the authors attributed the negative impact of growing wheat on wheat to the increased impact of leaf diseases.

For some cereal leaf spot diseases, the potential for seed-borne inoculum and seed-to-seedling transmission may negate the benefit of extended rotation intervals. For example, Turkington et al. (2006) found substantial levels of both scald and net blotch developed in fields not planted to barley or forage grasses for the previous 4 years. Although they did not measure seed-borne inoculum levels, in another study of seed-borne barley pathogens Turkington et al. (2002) found *Pyrenophora teres* (cause of net blotch) infected seed as much as 89%, with average annual levels of infection that ranged from 15% to 37%. In Alberta, Skoropad (1959), Lee et al. (1999) and Xi and Burnett (1997) found that *Rhynchosporium secalis*, the causal agent of scald, infected barley awns, lemmas, paleas and pericarps. The movement of scald into scald-free areas via infected seed has been suggested by a number of researchers (Habgood, 1971; Reed, 1957; Skoropad, 1959). Research in the 1950s and 1970s demonstrated transmission of *R. secalis* from infected seeds to seedlings mainly under greenhouse conditions (Habgood, 1971; Jackson and Webster, 1976; Kay and Owen, 1973; Reed, 1957; Skoropad, 1959). Reed (1957) also observed that infected seed that remained on the soil surface after seeding could act as a source of inoculum. In some countries, including India, New Zealand and England seed-borne inoculum of net blotch is considered to be an important source of disease (Hampton, 1980; Jordan, 1981; Sheridan et al., 1983; Shipton et al. 1973; Singh and Chand 1985).

Research from the United States has indicated that seed infection of wheat may be a factor to consider when using rotation as a means of cereal leaf spot management. In Florida, Luke et al. (1983) found decreased development of *Septoria nodorum* blotch, caused by *P. nodorum*, with either a 1 or 2 year rotation using uninfected seed, but not when *P. nodorum* infected seed was used. Similar results were also found by Milus and Chalkley (1997) in Arkansas where low (7%) and high (34–40%) levels of seed infection with *P. nodorum* led to significant leaf infection in an experiment planted in an area not cropped to wheat for at least the previous 4 years. They concluded that crop rotation should be combined with appropriate seed treatment to help prevent outbreaks of *Septoria nodorum* blotch. In New York State, Shah et al. (1995) found significant relationships between *P. nodorum* seed infection levels of up to 40% and subsequent development of *Septoria nodorum* blotch in winter wheat. Seed to seedling transmission of *P. tritici-repentis*, the causal agent of tan spot of wheat, has been demonstrated under greenhouse conditions and under laboratory conditions and in artificial potting media under outside conditions (Luz et al., 1998; Schilder and Bergstrom, 1995).

Published by Burleigh Dodds Science Publishing Limited, 2019.

3 Increasing spatial diversity: intercropping

Most research on intercropping has focused on agronomic factors and crop productivity (Boudreau and Mundt, 1997). However, intercropping of different crop species to increase the amount of diversity within an individual field can also improve disease management (Boudreau and Mundt, 1997; Finckh and Wolfe, 1998; Kantor, 1999; Sullivan, 2001; Wolfe, 2002). Four main types of intercropping systems have been identified:

- Mixed intercropping, where two or more crops are grown in the same field with no distinct planting pattern.
- Row intercropping, where two or more crops are grown in the same field with at least one of the components grown in distinct rows.
- Strip intercropping, where two or more crops are grown in the same field, but in strips that are wide enough to facilitate the use of farm equipment.
- Relay intercropping, where an additional crop is planted into an existing plant stand.

Burdon (1978) and Boudreau and Mundt (1997) have reviewed the ways intercropping can limit disease development. Because the proportion of susceptible host tissue decreases within the crop, intercropping reduces the production, amount and effectiveness of inoculum available for further spread and development within the crop. Secondly, increasing the space between susceptible hosts within crops also results in pathogen inoculum having to travel greater distances, impeding disease development. A third mechanism involves interception or filtering of pathogen propagules by the non-host component of the intercrop. As an example, Bannon and Cooke (1998) found that the clover in an intercrop with wheat reduced the dispersal of *Zymoseptoria tritici* (Desm.) Quaedvlieg & Crous pycnidiospores horizontally from an inoculum source or vertically from lower infected leaves. A fourth mechanism, cross-protection, may be similar to the induced resistance mechanism that has been suggested to occur for mixtures of the same crop species where non-virulent pathogen races trigger defence responses in the host, which then limit the impact of virulent races; however, cross-protection would involve non-host crops and non-pathogenic organisms (Boudreau and Mundt, 1997; Brown, 1975; Burdon, 1978). One other potential mechanism may occur via an influence on the microenvironment. Variability within the intercrop as a result of the presence of morphologically different crop components or an influence via an individual component of the intercrop canopy may produce less favourable microenvironmental conditions, leading to a reduction in disease development (Boudreau and Mundt, 1997; Burdon, 1978).

Hummel et al. (2009) studied the impact of intercropping wheat and canola on crop productivity and disease management. There was a negative relationship between wheat leaf spot severity (mainly the *Septoria* leaf spot complex) and increasing proportions of canola in the intercropping treatments. However, at one site, levels of disease were higher for the intercropped treatment versus wheat alone, while at the other site disease levels were similar. Overall, yield differences between intercropped and monoculture wheat were limited and Hummel et al. (2009) concluded there was insufficient benefit to recommend adoption of intercropping. Pridham and Entz (2008) found that intercropping of wheat with flax or field pea reduced leaf disease levels in wheat, but yields were variable and not consistently better for the intercropping treatments. Bulson et al. (1997) compared wheat only with varying proportions of wheat and field bean and found that intercropping increased the level of powdery mildew (*Blumeria graminis* (DC) Speer f. sp. *Tritici* emend.

Published by Burleigh Dodds Science Publishing Limited, 2019.

É. J. Marchal). Although yields were highest for the wheat alone sown at the recommended density, land equivalent ratios (LER) were higher with the intercrop, especially when wheat was sown at the recommended density of 75%. Bedoussac et al. (2008) found that durum wheat diseases were not reduced by growing an intercrop of durum and winter pea. More recently, Chapagain (2014) found that intercropping of wheat and common bean or faba bean did not influence the level of disease compared with wheat monoculture. In contrast, Zhu et al. (2017) found intercropping of wheat and faba bean reduced powdery mildew levels and improved yield.

In barley, Kinane and Lyngkaer (2002) detected significant reductions in net blotch severity only when barley was intercropped with field pea, while none of the intercropping treatments reduced leaf rust severity. Chapagain (2014) reported that intercropping of barley and field pea did not influence the level of disease compared with barley monoculture. Vilich-Meller (1992) found that intercropping of spring barley or oat reduced powdery mildew (*B. graminis* f.sp. *hordei*) in barley and leaf blotch (*Drechslera avenacea* (M.A. Curtis ex Cooke) Shoemaker) in oat. The impact of intercropping barley with a living mulch of kura clover was studied by Kosinski et al. (2011) in 2006 and 2007. In both years they found that spot-form net blotch severity was significantly reduced in rotations containing kura clover intercropped with barley versus barley alone. However, silage yields in 2006 were compromised by competition from the Kura clover in rotation compared with the barley only. In contrast, slight improvements in overall silage yield occurred with the intercropping treatment in 2007, although Kura clover made up the majority of the silage yield. Recently, a 9 year trial based on a 3 year rotational cycle was completed at two sites in Alberta, Canada, that assessed the impact of intercropping of barley variety mixtures, and variety rotation on silage production (Turkington et al., 2015a,b). For both locations leaf disease, primarily net-form net blotch, was generally highest for continuous production of the same barley variety, and lowest for mixtures or intercrops with barley, oat and triticale, especially where the variety components changed each year. Silage yields tended to be lowest for the continuous barley with the same variety, highest for the intercropping treatments with the same or different varieties, and intermediate for barley mixtures, especially when disease levels were increased. Overall results indicated that the addition of diversity in terms of different crop types and barley genetics can help to reduce the level of leaf disease and improve silage productivity.

4 Increasing genetic diversity: gene deployment

As noted earlier, there has been a reduction in genetic diversity in modern crop production in favour of uniform cultivars that meet a variety of agronomic, yield and quality requirements (Cowling, 1978). The development of a narrow range of modern high-yielding cultivars has been at the expense of genetic variability, presumably reducing resistance to plant pests (Harlan, 1972; Cowling, 1978; Wolfe, 2002). To address the problem of limited genetic diversity within our modern cropping systems, a number of authors have recommended the use of multilines, variety mixtures, gene deployment as well as the use of intercropping (Apple, 1977; Boudreau and Mundt, 1997; Browning and Frey, 1969; Cowling, 1978; Cox et al., 2004; Finckh et al., 2000; Finckh and Wolfe, 1998; Frey et al., 1977; Fry, 1982; Garrett and Mundt, 1999; Harlan, 1972; McDonald and Linde, 2002a,b; Newton et al., 2009; Priestly, 1981; Wolfe, 1985, 2002; Zadoks and Schein, 1979).

Gene deployment has been suggested as a way to add diversity to our modern cropping systems (Adugna, 2004; Burdon et al., 2014; Finckh and Wolfe, 1998; Frey et al., 1977; Fry, 1982; Priestly, 1981; McDonald and Linde, 2002a,b; REX Consortium, 2016; Zhan et al., 2015). Gene deployment can be achieved via spatial deployment locally or regionally or gene rotation over time and may involve near-isogenic lines or the use of diverse crop varieties with different resistance genes. For example, resistance genes deployed in one area may be effective against pathogen races originating from another region where crop varieties with different resistance genes are being grown. Gene deployment has been suggested as a strategy to increase the durability of resistance genes and reduce the spread and development of cereal rusts in North America (Frey et al., 1973; Fry, 1982). On a smaller spatial scale, Priestley (1981) suggested that disease spread can be restricted if neighbouring fields are planted to cultivars with different resistance genes. Priestley (1981) and Priestly and Wolfe (1977) developed diversification schemes to assist farmers with on-farm deployment of barley and winter wheat cultivars among farms within a single year and in successive years to reduce the risk of powdery mildew and stripe rust, respectively.

Temporal deployment of cultivars with different resistance genes (gene rotation) may also be a potential method of adding a small level of diversity to a monoculture system (Finckh and Wolfe, 1998), where crop rotation may not be a viable option. For example, diversity in time with regard to different crop species for an individual field may be difficult if not impossible to implement for livestock operations, given on-farm feed requirements and market factors. Continuous barley production leads to a build-up of disease in these fields and a general reduction in yield potential over the long-term (Mathre, 1997; Tekauz, 2003). One option that farmers may use to help reduce leaf diseases, while maintaining productivity, is to introduce some level of diversity by deploying different varieties in subsequent growing seasons. By changing varieties, the farmer has the potential to change the disease resistance genes in use each year, thereby reducing the impact of disease. For example, Turkington et al. (2005) demonstrated that growing the same barley variety for 3 years in a row resulted in significantly higher scald and net blotch severity and lower yield and grain size compared to barley variety rotation. However, reductions in disease and improvements in yield and grain size were highest when barley was grown on triticale versus barley on barley residue.

5 The role of conservation tillage

The use of tillage as a form of sanitation has been advocated as a traditional method of cereal leaf disease control (Maloy, 2005; Mathre, 1997; Menzies and Gilbert, 2003; Palti, 1981; Tekauz, 2003; Turkington, 2003; Wiese, 1987). Tillage may be extensive, for example mouldboard ploughing or the use of a disker followed by several cultivator passes and harrowing to prepare the soil for seeding as well as weed management (Daigh and DeJong-Hughes, 2017; DeJong-Hughes and Daigh, 2017a). Alternatively, tillage may be less aggressive and accomplished using chisel ploughs/shovel-equipped cultivators with several passes over the field followed by harrowing and seeding. The combination of breaking up crop residues into smaller pieces coupled with mixing and burial of crop residues in the soil can help to accelerate decomposition of pathogen-infected crop residues, thus removing a potential source of inoculum. Tillage also places infected crop residues beneath a layer of soil where pathogen fruiting bodies and/or spores are unable to be dispersed and infect above-ground plant parts (Palti, 1981). Unfortunately, tillage can have significant negative effects on soil health and sustainability and promotes water evaporation (Daigh and DeJong-Hughes, 2017; DeJong-Hughes and Daigh, 2017b; Derpsch and Freidrich, 2009; Derpsch, 2015; Friedrich et al., 2012; Soane et al., 2012).

As a result there has been growing interest in and implementation of conservation tillage for crop production (Awada et al., 2014; Derpsch, 2015; Derpsch and Friedrich, 2009; Kassam et al., 2015). Conservation tillage is seen as a method of ensuring environmentally sustainable agricultural production that enhances soil quality, prevents erosion, promotes soil biological activity and diversity, and reduces labour and fuel costs (Awada et al., 2014; Busari et al., 2015; Daigh and DeJong-Hughes, 2017; DeJong-Hughes and Daigh, 2017b; Derpsch, 2015; FAO, 2015; Friedrich et al., 2012; Soane et al., 2012). In contrast, conventional tillage represents a significant disruption to the soil and soil surface environments where the stabilizing characteristics and mechanisms that function in natural ecosystems are greatly impaired. Unfortunately, the direct impact of conservation tillage is the retention of crop residues and pathogen propagules at or just below the soil surface, which may then act as an inoculum source, thus increasing the risk of disease (Awada et al., 2014; Coutts and Smith, 1991; Evans and Fleury, 1993; Leduc, 1994; Watkins and Boosalis, 1994). An overview of the potential impact of conservation versus conventional tillage on cereal leaf diseases will be presented.

Under conservation tillage, moisture levels are increased at the soil surface (Steiner, 1994), the rate of evaporation is reduced (Sadler and Turner, 1994; Steiner, 1994) and there is a general reduction in temperature and dampening of daily fluctuations (Clegg and Francis, 1994; Horton et al., 1994). Increased moisture levels and cooler temperatures close to the soil surface may increase disease risk (Rothrock, 1992). Alternatively, there may be increased population levels, activity and diversity of various organisms in the soil surface environment and this may help to reduce the risk of disease (Cochran et al., 1994; Cook and Baker, 1983; Cook and Veseth, 1991; Lupwayi et al., 1998). Stubble retention on the soil surface may decrease the amount of photosynthetically active radiation at the soil surface as a result of shading, but residue will also offer physical protection of the surface helping to prevent wind and water erosion (Alberts and Neibling, 1994; Fryrear and Bilbro, 1994; Wilkins et al., 1988). As Watkins and Boosalis (1994) have suggested, retention of stubble on the soil surface may influence overall plant health and this may either enhance or restrict the plant's ability to withstand pathogen attack.

In general, the influence of conservation tillage on above-ground plant diseases may be fairly obvious; the more infected crop residue on the soil surface the greater the disease risk (Watkins and Boosalis, 1994). Residue-borne barley and wheat leaf disease pathogens sporulate on infected plant residues and these spores then serve as primary sources of inoculum (Mathre, 1997; Menzies and Gilbert, 2003; Tekauz, 2003; Wiese, 1987). Because there is an abundant reservoir of infected residue on the soil surface under conservation tillage, disease may appear earlier and subsequent development may be accelerated producing greater disease levels at crop maturity. Pittelkow et al. (2015) conducted a meta-analysis using a global dataset and while not directly comparing no-till and conventional tillage in terms of plant disease levels, they found an overall 5.7% reduction in yield under no-till versus conventional tillage. However, when they focused on dry land crop production in semi-arid regions, no-till resulted in a 7.3% increase in yield when it was combined with residue retention and crop rotation and they attributed this to 'improved water infiltration and greater soil moisture conservation'. Research investigating the impact of conservation tillage on cereal leaf diseases has been conducted since the mid- to late 1980s in the prairies of western Canada, where there has been extensive adoption of conservation tillage starting in the 1990s. Traditionally in western Canada the most common form of conventional tillage is the use of a heavy duty cultivator (chisel plough); mouldboard ploughing is uncommon (Green, 1984). In western Canada, typical conventional tillage practices leave between 50% and 80% residue cover after each tillage operation, depending on the tillage implement used (Jensen and Timmermans, 1991; MAFRI, 2008), but these residues still act as a significant source of pathogen inoculum. Although ploughing leaves behind a smaller amount of residue, subsequent tillage operations related to seedbed preparation likely bring infected crop residues back to the soil surface.

From 1994 to 2001, Bailey et al. (2000) and Kutcher et al. (2011) studied the impact of various crop and pest management practices on managing leaf diseases in cereals. Tillage treatments included conventional tillage with tillage operations in the previous fall and the following spring before seeding; minimum tillage with only spring tillage prior to seeding; and zero tillage with no tillage operations before seeding. Levels of net blotch in barley and a combination of tan spot and the *Septoria* leaf spot complex in wheat were similar for all tillage regimes, while barley and wheat yields were similar amongst tillage systems in most years except barley in 1994 and 2001 where yields were significantly higher for zero tillage versus conventional tillage. The latter year was very dry with low severity of leaf spot diseases and the yield benefit of zero tillage was attributed to the increase in soil moisture under zero tillage. In the latter study, Kutcher et al. (2011) detected similar levels of the leaf spot complex in wheat among tillage systems.

The lack of a tillage effect on cereal leaf diseases was also previously reported by Bailey et al. (1992), where the tillage system did not significantly affect tan spot severity in 3 of 4 years of the trial. In 1992 and 1993, Bailey (1996) reported slightly higher overall leaf spot severity under conventional tillage versus zero tillage, while isolation of the causal agents of speckled leaf blotch (*S. tritici*) and tan spot (*P. tritici-repentis*) was slightly higher under zero tillage. In contrast no differences in isolation of the causal agent of leaf/glume blotch (*P. nodorum*) occurred, while isolation of the spot blotch fungus (*Bipolaris sorokiniana* (*Sacc.*) Shoemaker) was lower overall, and was slightly lower under zero tillage. From 1991 to 1998, Bailey et al. (2001) found wheat leaf spot levels and grain yields were similar under zero, minimum and conventional tillage. In a survey of commercial wheat fields in Saskatchewan, Fernandez et al. (2009) reported that there was limited impact of tillage system on overall leaf spot severity, but there were effects on the percentage isolation of causal agents. Isolation of the causal agent of tan spot increased under conservation tillage, while isolation of the spot blotch pathogen was increased under conventional tillage, with isolation of the causal agents of the *Septoria* leaf spot complex reduced under zero tillage. Similar results were found in Manitoba by Gilbert and Woods (2001) where under conservation tillage there was a decrease in isolation of the causal agents of the *Septoria* leaf spot complex, but an increase in isolation of the tan spot fungus compared with conventional tillage. In a survey of commercial barley fields in 1995–7 in Alberta, Turkington et al. (2006) reported that the probability of higher levels of scald or net blotch in fields under conventional, minimum or zero tillage was similar regardless of the tillage system used.

In earlier work by Jordan and Allen (1984) in the United Kingdom, higher net blotch inoculum loads, disease levels and earlier disease development were reported when crop residues remained on the soil surface versus ploughing of crop residues into the soil. Bockus and Claassen (1992) found that the ploughing under of crop residues in a continuous wheat rotation significantly reduced tan spot levels compared with no till, with intermediate disease levels for the chisel plough treatment. Wheat yield was significantly higher in 2 of 4 years for the plough treatment compared with no till, while yields for the chisel plough treatment were significantly higher than no till in 1 of 4 years. In contrast, under a wheat–sorghum rotation differences in tan spot levels among tillage treatments were less consistent, while yields were similar among tillage regimes. In a study not directly comparing conventional and conservation tillage, Adee and Pfender (1989) suggested that the level of primary inoculum, that is tan spot infected crop residues, was related to disease development, and management of infected residues was recommended to reduce tan spot development and impact. Rees et al. (1982) also did not directly compare conservation tillage and conventional tillage regimes in Australia, but did demonstrate that tan spot levels and percentage yield loss increased as the amount of infected residue applied to plot areas increased. Schuh (1990) compared tan spot development in four commercial winter wheat fields, two under conventional tillage (i.e. mouldboard ploughing followed by tillage to prepare the seedbed) and two under conservation tillage (disking followed by chisel ploughing). In general, disease levels were lower for the conventional versus conservation tillage treatments. Jørgensen and Olsen (2007) also found that reduced tillage with only harrowing resulted in significantly higher levels of tan spot in winter wheat planted into winter wheat residues, while *Septoria* leaf spot and powdery mildew were lower compared with the ploughed treatment. In both years of their study in Denmark, winter wheat yields were higher for the ploughed treatment versus reduced tillage.

In the Northern Great Plains of the United States, Krupinsky et al. (2007) evaluated the effect of tillage and nitrogen regimes on wheat leaf spot severity (tan spot and the *Septoria* complex) from 1986 to 1996. Overall, no till/direct seeding had limited to no impact on leaf spot severity, especially during dry years, 1988–92. However, no till/ direct seeding did result in higher leaf spot severity in wet years, but only in 15 of 47 evaluations, 9 of 61 evaluations and 7 of 50 evaluations, for the spring wheat annual, spring wheat fallow and the winter wheat sunflower rotations, respectively. Furthermore, there were interactions between tillage regime and nitrogen fertilizer rate, whereby no till/direct seeding had higher leaf spot levels compared with conventional and minimum tillage regimes when the nitrogen rate was low, but at higher nitrogen rates disease levels were similar. In Poland, Sawinska et al. (2006) found that levels of powdery mildew, tan spot and *Septoria* leaf blight in winter wheat were increased under conventional tillage versus direct seeding, especially with adequate moisture. Andert et al. (2016) investigated the impact of crop rotation and tillage on the intensity of fungicide use in Germany for a range of crops and found that ploughing, compared to not ploughing, reduced fungicide intensity in winter types of wheat, barley and rye. Carignano et al. (2008) investigated the impact of tillage/residue management (no till, reduced till and burning of residue), fungicide application, nitrogen fertility and cultivar on leaf spot severity and productivity in a winter wheat on winter wheat rotation over 2 years at a total of five site years in Kansas. They found a significant increase in leaf spot severity under no till, while the difference in yield between no till and reduced till, although significant, was relatively small. Significant interactions were detected between tillage regime, and cultivar or fungicide. In general, differences among tillage systems were greatest when no fungicide was applied and a susceptible cultivar was grown, while disease levels and productivity were similar among tillage systems when either a fungicide or resistant cultivar was used.

6 Conclusions and future trends

As has been noted, modern agro-ecosystems are often composed of either monocultures or rotations with a small number of crop components and uniform cultivars, resulting in reduced genetic diversity and a narrow genetic base which compromises disease resistance (Cowling, 1978; Harlan, 1972; Wolfe, 2002; Zadoks and Schein, 1979). These simplified agro-ecosystems are vulnerable to abiotic and biotic stresses, and often allow pest species which can exploit these systems to proliferate rapidly (Finckh and Wolfe, 1997; Tanaka et al., 2002; Wolfe, 2002). The use of techniques such as variety mixtures, intercropping, rotation and gene deployment can help to increase genetic diversity, particularly if combined (Chaube and Singh, 1991).

On-farm implementation of techniques to enhance diversity need not be expensive or cumbersome. For example, the addition of diversity for plant disease management within a forage production system can be achieved through the use of mixtures or intercropping. Agronomically adapted and superior crop cultivars can be combined to capitalize on the benefits of adding diversity to this type of cropping system. Unfortunately, the potential use of diversity via mixtures or intercropping may be more problematic, especially in the short term in agro-ecosystems focused on food production where end users are more concerned about uniformity and specific commodity characteristics (Finckh et al., 2000; Mundt, 2002; Newton et al., 1998; Newton and Swanston, 1999; Wolfe, 2000, 2002). However, close collaboration among researchers, farmers and commodity end users, coupled with the use of crop components that are agronomically compatible and that have complementary quality characteristics, may help to illustrate the potential of genetic diversity in the form of intercropping.

Crop rotation still has a significant role to play in adding diversity to current and future cropping systems. However, crop rotation will always need to be considered within the context of the myriad other factors that influence a farmer's crop choice and land use decisions (Cook and Veseth, 1991; Karlen et al., 1994; Tanaka et al., 2002). More than 80 years ago, Howard and Stakman (1921a) indicated that rotations need to be based on crops that are profitable for the farmer, otherwise the benefit in terms of disease management is negated. Nevertheless farmers can make more informed decisions about the use of rotations when trying to balance competing goals in terms of production, economic and sustainability requirements. Parker (1915) in his book *Field Management and Crop Rotation* outlined the important role of crop rotation and this quote still applies today:

> Crop rotation in itself is not the cure-all for unproductive land or the absolute key to profits from high priced agricultural land. But crop rotation is the chief factor in a combination of good farming practices that will maintain the productivity of the soil, and around which intensive systems of farming may be developed that will yield the maximum crop value per acre at the minimum of expense. Crop rotation is to general field agriculture what the foundation is to the house, the solid base on which we may successfully rear a permanent superstructure designed in a hundred different ways according to our individual requirements and desires.

For farmers to truly capitalize on the potential benefits of using diversity for disease management, they need to have a good understanding of the life cycle of the pathogen(s) they are trying to manage. The importance of knowledge of the plant

Published by Burleigh Dodds Science Publishing Limited, 2019.

disease issues that challenge farmers is not a new concept. In the book *Farm Economy: A Cyclopedia of Agriculture for the Practical Farmer and His Family*, Howard and Stakman (1921b) stressed the importance of knowledge regarding the pathogen(s) of concern in the following quote:

> It therefore becomes necessary to know what kind of germ is causing a particular disease, and further to know the habits of this particular germ. This is all the more true, since the different kinds of germs may have very different habits, and therefore must be controlled by different methods. It is absolutely necessary therefore for a farmer, before he attempts to control a disease, to know what germ causes the disease and how it acts in order that he may apply his preventative measures at the proper time and in the proper place.

Knowledge regarding how a pathogen survives and develops will illustrate where strategies in addition to genetic diversity and rotation need to be used. Moreover, farmers are typically faced with more than one disease, thus an integrated approach will be needed; one that is based on a diversity of disease and crop management strategies.

The retention of crop residues at or just below the soil surface may have both a direct and an indirect influence on the development of cereal leaf diseases (Watkins and Boosalis, 1994). Overall, crop production factors such as environmental variation among individual fields, regions and years, crop rotation, seed-borne inoculum and choice of cultivar (level of disease resistance) will likely have a larger impact on the risk of disease development compared with the type of tillage system that is used (Bailey, 1996; Bailey and Duczek, 1996; Krupinsky et al., 2002). Overall, there is potential for residue-borne cereal leaf diseases to be effectively managed under conservation tillage by the use of sound rotations and agronomic practices, and careful choice of variety.

7 Where to look for further information

The References section contains key publications and websites that provide introductions and more in-depth overviews of subject areas related to crop rotation, conservation tillage and intercropping as well as their impact in relation to cereal leaf spot diseases.

Additional useful sources of information include:

- http://asi.ucdavis.edu/programs/sarep/about/copy_of_what-is-sustainable-agriculture/practices/conservation-tillage
- http://www.fao.org/docrep/012/al298e/al298e.pdf
- http://www.fao.org/conservation-agriculture/overview/principles-of-ca/en/
- http://publications.gov.sk.ca/documents/20/85517-principle%20practices%20crop%20rotation%202017.pdf
- http://apsnet.org/edcenter/Pages/default.aspx
- http://www.agr.gc.ca/eng/science-and-innovation/agricultural-practices/soil-and-land/soil-management/issues-management-problems-and-solutions-for-maintaining-a-zero-tillage-system-and-other-beneficial-soil-management-practices-1of10/?id=1238006852354
- https://grdc.com.au/resources-and-publications

Published by Burleigh Dodds Science Publishing Limited, 2019.

8 References

Adee, E. A. and Pfender, W. F. (1989), The effect of primary inoculum level of *Pyrenophora tritici-repentis* on tan spot epidemic development in wheat, *Phytopathology*, 79, 873–7.

Adugna, A. (2004), Alternative approaches in deploying genes for disease resistance in crops, *Asian J. Plant Sci.*, 3, 618–23.

Alberts, E. E. and Neibling, W. H. (1994), Influence of crop residues on water erosion, in P. W. Unger (Ed.), *Managing Agricultural Residues*, CRC Press, Inc., Boca Raton, FL, pp. 19–39.

Andert, S., Bürger, J., Stein, S. and Gerowitt, B. (2016), The influence of crop sequence on fungicide and herbicide use intensities in north German arable farming, *Eur. J. Agron.*, 77, 81–9.

Angus, J. F., Kirkegaard, J. A., Hunt, J. R., Ryan, M. H., Ohlander, C. L. and Peoples, M. B. (2015), Break crops and rotations for wheat, *Crop Pasture Sci.*, 66, 523–52.

Apple, J. L. (1977), The theory of disease management, in J. G. Horsfall and E. B. Cowling (Eds), *Plant Disease an Advance Treatise: How Disease is Managed*, Volume 1, Academic Press Inc., New York, NY, pp. 79–101.

Awada, L., Lindwall, C. W. and Sonntag, B. (2014), Conservation tillage impacts on soil, crop and the environment, *Int. Soil Water Conserv. Res.*, 2, 47–65.

Aylor, D. E. (1990), The role of intermittent wind in the dispersal of fungal pathogens, *Annu. Rev. Phytopathol.*, 28, 73–92.

Aylor, D. E. (2003), Spread of plant disease on a continental scale: Role of aerial dispersal of pathogens, *Ecology*, 84, 1989–97.

Bailey, K. L. (1996), Diseases under conservation tillage systems, *Can. J. Plant Sci.*, 76, 635–9.

Bailey, K. L. and Duczek, L. J. (1996), Managing cereal diseases under reduced tillage, *Can. J. Plant Pathol.*, 18, 159–67.

Bailey, K. L., Mortensen, K. and Lafond, G. P. (1992), Effects of tillage systems and crop rotations on root and foliar diseases of wheat, flax, and peas in Saskatchewan, *Can. J. Plant Sci.*, 72, 583–91.

Bailey, K. L., Johnston, A. M., Kutcher, H. R., Gossen, B. D. and Morrall, R. A. A. (2000), Managing crop losses from foliar diseases with fungicides, rotation, and tillage in the Saskatchewan Parkland, *Can. J. Plant Sci.*, 80, 169–75.

Bailey, K. L., Gossen, B. D., Lafond, G. P., Watson, P. R. and Derksen, D. A. (2001), Effect of tillage and crop rotation on root and foliar diseases of wheat and pea in Saskatchewan from 1991 to 1998: Univariate and multivariate analyses, *Can. J. Plant Sci.*, 81, 789–803.

Bankina, B., Bimšteine, G., Ruža, A., Kreita, D., Katamadze, M. and Paura, L. (2015), Crop rotation – The main factor influencing the development of wheat leaf blotch, in *Proceedings of the 25th NJF Congress*, Riga, Latvia, 16–18 June 2015, pp. 65–9.

Bankina, B., Bimšteine, G., Arhipova I., Kaņeps, J. and Stanka, T. (2018), Importance of agronomic practice on the control of wheat leaf diseases, *Agriculture*, 8, 56. https://doi.org/10.3390/agriculture8040056 (accessed 10 June 2018).

Bannon, F. J. and Cooke, B. M. (1998), Studies on dispersal of *Septoria tritici* pycnidiospores in wheat–clover intercrops, *Plant Pathol.*, 47, 49–56.

Barnes, E. H. (1964), Changing plant disease losses in a changing agriculture, *Phytopathology*, 54, 1314–19.

Bedoussac, L., Matura, M. and Hemptinne, J.-L. and Justes, E. (2008), Is durum wheat-winter pea intercropping efficient to reduce pests and diseases? in *10th Congress of European Society for Agronomy*, 15–19 September 2008, Bologna, Italy.

Black, A. L., Siddoway, F. H. and Brown, P. L. (1974), Summer fallow in the northern Great Plains (winter wheat), in *Summer Fallow in the Western United States*, USDA-ARS Conserv. Res. Rep. no. 17, USDA, Washington DC, pp. 36–50.

Bockus, W. W. and Claassen, M. M. (1992), Effects of crop rotation and residue management practices on severity of tan spot of winter wheat, *Plant Dis.*, 76, 633–6.

Boudreau, M. A. and Mundt, C. C. (1997), Ecolgical approaches to disease control, in N. A. Rechcigl and J. E. Rechcigl (Eds), *Environmentally Safe Approaches to Crop Disease Control*, CRC Press LLC, Boca Raton, FL, pp. 33–62.

Brown, J. F. (1975), Factors affecting the relative ability of strains of fungal pathogens to survive in populations, *J. Aust. Inst. Agric. Sci.*, 41, 3–11.

Brown, J. K. M. and Hovmøller, M. S. (2002), Aerial dispersal of pathogens on the global and continental scales and its impact on plant disease, *Science*, 297, 537–41.

Browning, J. A. and Frey, K. J. (1969), Multiline cultivars as a means of disease control, *Annu. Rev. Phytopathol.*, 7, 355–82.

Bulson, H. A. J., Snaydon, R. W. and Stopes, C. E. (1997), Effects of plant density on intercropped wheat and field beans in an organic farming system, *J. Agric. Sci.*, 128, 59–71.

Burdon, J. J. (1978), Mechanisms of disease control in heterogeneous plant populations – an ecologist's view, in P. R. Scott and A. Bainbridge (Eds) *Plant Disease Epidemiology*, Blackwell Science Publishing, Oxford, UK, pp. 193–200.

Burdon, J. J., Barrett, L. G., Rebetzke, G. and Thrall, P. H. (2014), Guiding deployment of resistance in cereals using evolutionary principles, *Evol. Appl.*, 7, 609–24.

Burgess, L. W. and Griffin, D. M. (1968), The recovery of *Gibberella zeae* from wheat straws, *Aust. J. Exp. Agric. Anim. Husb.*, 8, 364–70.

Busari, M. A., Kukal, S. S., Kaur, A., Bhatt, R. and Dulazi, A. A. (2015), Conservation tillage impacts on soil, crop and the environment, *Int. Soil Water Conserv. Res.*, 3, 119–29.

Carignano, M., Staggenborg, S. A. and Shroyer, J. P. (2008), Management practices to minimize tan spot in a continuous wheat rotation, *Agron. J.*, 100, 145–53.

Chapagain, T. (2014), Intercropping wheat and barley with nitrogen fixing legume species in low input organic systems, Ph.D. Thesis, Faculty of Graduate and Postdoctoral Studies, University of British Columbia, 192pp.

Chaube, H. S. and Singh, U. S. (1991), *Plant Disease Management: Principles and Practices*, CRC Press, Boca Raton, FL.

Chen, X. M. (2005), Epidemiology and control of stripe rust [*Puccinia striiformis* f. sp. *tritici*] on wheat, *Can. J. Plant Pathol.*, 27, 314–37.

Clegg, M. D. and Francis, C. A. (1994), Crop management, in J. L. Hatfield and D. L. Karlen (Eds), *Sustainable Agriculture Systems*, CRC Press, Inc., Boca Raton, FL, pp. 135–56.

Cochran, V. L., Sparrow, S. D. and Sparrow, E. B. (1994), Residue effects on soil micro- and macro-organisms, in P. W. Unger (Ed.), *Managing Agricultural Residues*, CRC Press, Inc., Boca Raton, FL, pp. 163–84.

Cook, R. J. (2000), Advances in plant health management in the twentieth century, *Annu. Rev. Phytopathol.*, 38, 95–116.

Cook, R. J. and Baker, K. F. (1983), *The Nature and Practice of Biological Control of Plant Pathogens*, APS Press, St Paul, MN, 539pp.

Cook, R. J. and Veseth, R. J. (1991), *Wheat Health Management. Plant Health Management Series*, APS Press, St Paul, MN, 152pp.

Coutts, G. R. and Smith, R. K. (1991), *Zero Tillage Production Manual*, Manitoba North Dakota Zero Tillage Farmers Association, Brandon, Manitoba, 42pp.

Cowling, E. B. (1978), Agricultural and forest practices that favor epidemics, in J. G. Horsfall and E. B. Cowling (Eds), *Plant Disease an Advance Treatise: How Disease Develops in Populations*, Volume 2, Academic Press Inc., New York, NY, pp. 361–81.

Cox, C. M., Garrett, K. A., Bowden, R. L., Fritz, A. K., Dendy, S. P. and Heer, W. F. (2004), Cultivar mixtures for the simultaneous management of multiple diseases: Tan spot and leaf rust of wheat, *Phytopathology*, 94, 961–9.

Daigh, A. and DeJong-Hughes, J. (2017), A brief history of soil tillage and tillage research, *Upper Midwest Tillage Guide: Tillage Guide Part 1*, University of Minnesota Extension, https://www.extension.umn.edu/agriculture/soils/tillage/tillage-guide-history/docs/upper-midwest-tillage-guide-history.pdf (accessed 9 March 2018).

DeJong-Hughes, J. and Daigh, A. (2017a), Tillage implements, purpose and ideal use, *Upper Midwest Tillage Guide: Tillage Guide Part 2*, University of Minnesota Extension, https://www.extension.umn.edu/agriculture/soils/tillage/tillage-guide-implements/docs/upper-midwest-tillage-guide-implements.pdf (accessed 9 March 2018).

DeJong-Hughes, J. and Daigh, A. (2017b), Reducing tillage intensity, *Upper Midwest Tillage Guide: Tillage Guide Part 3*, University of Minnesota Extension, https://www.extension.umn.edu/agriculture/soils/tillage/tillage-guide-intensity/docs/upper-midwest-tillage-guide-intensity.pdf (accessed 9 March 2018).

Derpsch, R. (2015), Frontiers in conservation tillage and advances in conservation tillage, Food and Agriculture Organization of the United Nations, Agriculture and Consumer Protection Department, Conservation Agriculture, http://www.fao.org/ag/ca/6b.html (accessed 9 March 2018).

Derpsch, R. and Friedrich, T. (2009), Global overview of conservation agriculture adoption, *IV World Congress on Conservation Agriculture*, New Delhi, India, February 2009, Food and Agriculture Organization of the United Nations, Agriculture and Consumer Protection Department, Conservation Agriculture, CA adoption worldwide, http://www.fao.org/ag/ca/6b.html and http://www.fao.org/ag/ca/doc/Derpsch-Friedrich-Global-overview-CA-adoption3.pdf (accessed 9 March 2018).

Duczek, L. J., Sutherland, K. A., Reed, S. L., Bailey, K. L. and Lafond, G. P. (1999), Survival of leaf spot pathogens on crop residues of wheat and barley in Saskatchewan, *Can. J. Plant Pathol.*, 21, 165–73.

Evans, R. and Fleury, D. (1993), *Conservation Farming Guide*, Alberta Conservation Tillage Society, October 1993, 48pp.

Eversmeyer, M. G. and Kramer, C. L. (2000), Epidemiology of wheat leaf rust and stem rust in the central Great Plains of the USA, *Annu. Rev. Phytopathol.*, 38, 491–513.

Eversmeyer, M. G., Kramer C. L. and Browder, L. E. (1984), Presence, viability, and movement of *Puccinia recondita* and *P. graminis* inoculum in the Great Plains, *Plant Dis.*, 68, 392–95.

FAO (2015), Conservation agriculture, Food and Agriculture Organization of the United Nations. Agriculture and Consumer Protection Department, http://www.fao.org/ag/ca/index.html (accessed 9 March 2018).

Farr, D. F., Bills, G. F., Chamuris, G. P. and Rossman, A. Y. (1989), *Fungi on Plants and Plant Products in the United States*, American Phytopathology Society Press, St Paul, MN, 1252pp.

Fernandez, M. R., Zentner, R. P., McConkey, B. G. and Campbell, C. A. (1998), Effects of crop rotations and fertilizer management on leaf spotting diseases of spring wheat in southwestern Saskatchewan, *Can. J. Plant Sci.*, 78, 489–96.

Fernandez, M. R., Pearse, P. G., Holzgang, G., Basnyat, P. and Zentner, R. P. (2009), Impacts of agronomic practices on the leaf spotting complex of common wheat in eastern Saskatchewan, *Can. J. Plant Sci.*, 89, 717–30.

Fernandez, M. R., Wang, H., Cutforth, H. and Lemke, R. (2016), Climatic and agronomic effects on leaf spots of spring wheat in the western Canadian prairies, *Can. J. Plant Sci.*, 96, 895–907.

Finckh, M. R. and Wolfe, M. S. (1997), The use of biodiversity to restrict plant diseases and some consequences for farmers and society, *in* L. E. Jackson (Ed.), *Ecology in Agriculture*, Academic Press, San Diego, CA, pp. 199–233.

Finckh, M. R. and Wolfe, M. S. (1998), Diversification strategies, *in* D. G. Jones (Ed.), *The Epidemiology of Plant Diseases*, Chapman and Hall, London, UK, pp. 231–59.

Finckh, M. R., Gacek, E. S., Goyeau, H., Lannou, C., Merz, U., Mundt, C. C., Munk, L., Nadziak, J., Newton, A. C., de Vallavieille-Pope, C. and Wolfe, M. S. (2000), Cereal variety and species mixtures in practice, with emphasis on disease resistance, *Agronomie*, 20, 813–37.

Frey, K. J., Browning, J. A. and Simons, M. D. (1973), Management of host resistance genes to control disease loss, *Z. Pflanzenkr. Pflanzenschutz*, 80, 160–80.

Frey, K. J., Browning, J. A. and Simons, M. D. (1977), Management systems for host genes to control disease loss, *Ann. N.Y. Acad. Sci.*, 287, 255–74.

Friedrich, T., Derpsch, R. and Kassam, A. (2012), Overview of the Global Spread of Conservation Agriculture, *Field Actions Science Reports* [Online], Special Issue 6 | 2012, Online since 06 November 2012, connection on 18 June 2018. URL: http://journals.openedition.org/factsreports/1941 (accessed 18 June 2018).

Fry, W. E. (1982), *Principles of Plant Disease Management*, Academic Press, New York, NY.

Fryrear, D. W. and Bilbro, J. D. (1994), Wind erosion control with residues and related practices, in P. W. Unger (Ed.), *Managing Agricultural Residues*, CRC Press, Inc., Boca Raton, FL, pp. 7–17.

Garrett, K. A. and Mundt, C. C. (1999), Epidemiology in mixed host populations, *Phytopathology*, 89, 984–90.

Gilbert, J. and Woods, S. M. (2001), Leaf spot diseases of spring wheat in southern Manitoba farm fields under conventional and conservation tillage, *Can. J. Plant Sci.*, 81, 551–9.

Ginns, J. H. (1986), Compendium of plant disease and decay fungi in Canada, 1960–1980, Publication 1813, Research Branch, Agriculture Canada.

Green, F. M. (1984), Machinery for tillage and planting, Agdex 740-2, May 1984, Alberta Agriculture, Edmonton, Alberta.

Habgood, R. M. (1971), The transmission of *Rhynchosporium secalis* by infected barley seed, *Plant Pathol.*, 20, 80–1.

Hampton, J. G. (1980), The role of seed-borne inoculum in the epidemiology of net blotch of barley in New Zealand, *N. Z. J. Exp. Agric.*, 8, 297–9.

Harker, K. N., O'Donovan, J. T., Turkington, T. K., Blackshaw, R. E., Lupwayi, N. Z, Smith, E. G., Johnson, E. N., Gan, Y., Kutcher, H. R., Dosdall, L. M. and Peng, G. (2015), Canola rotation frequency impacts canola yield and associated pests, *Can. J. Plant Sci.*, 95, 9–20.

Harlan, J. R. (1972), Genetics of disaster, *J. Environ. Quality*, 1, 212–15.

Horton, R., Kluitenberg, G. J. and Bristow, K. L. (1994), Surface crop residue effects on the soil surface energy balance, in P. W. Unger (Ed.), *Managing Agricultural Residues*, CRC Press, Inc., Boca Raton, FL, pp. 143–62.

Howard, C. W. and Stakman, E. C. (1921a), Rotation as a means of blight-control, in *Farm Economy: A Cyclopedia of Agriculture for the Practical Farmer and his Family*, Better Farming Association, Minneapolis, MN, pp. 41–4.

Howard, C. W. and Stakman, E. C. (1921b), Diseases of cereal crops, in *Farm Economy: A Cyclopedia of Agriculture for the Practical Farmer and his Family*, Better Farming Association, Minneapolis, MN, pp. 762–76.

Hummel, J. D., Dosdall, L. M., Clayton, G. W., Turkington, T. K., Lupwayi, N. Z., Harker, K. N. and O'Donovan, J. T. (2009), Canola–wheat intercrops for improved agronomic performance and integrated pest management, *Agron. J.*, 101, 1190–7.

Jackson, L. F. and Webster, R. K. (1976), Seed and grasses as possible sources of *Rhynchosporium secalis* for barley in California, *Plant Dis. Reptr.*, 60, 233–6.

Jensen, T. and Timmermans, J. (1991), Conservation tillage, Agdex 516-3, September 1991, Agri-fax, Alberta Agriculture, Edmonton, Alberta.

Johnston, A. M., Kutcher, H. R. and Bailey, K. L. (2005), Impact of crop sequence decisions in the Saskatchewan Parkland, *Can. J. Plant Sci.*, 85, 95–102.

Jordan, V. W. L. (1981), Aetiology of barley net blotch caused by *Pyrenophora teres* and some effects on yield, *Plant Pathol.*, 30, 77–87.

Jordan, V. W. L. and Allen, E. C. (1984), Barley net blotch: Influence of straw disposal and cultivation methods on inoculum potential, and on incidence and severity of autumn disease, *Plant Pathol.*, 33, 547–59.

Jørgensen, L. N. and Olsen, L. V. (2007). Control of tan spot (*Drechslera tritici-repentis*) using cultivar resistance, tillage methods and fungicides, *Crop Prot.*, 26, 1606–16.

Kantor, S. (1999), Intercropping, agriculture and natural resources, Fact Sheet #531, Cooperative Extension, Washington State University, Renton, WA.

Karlen, D. L., Varvel, G. E., Bullock, D. G. and Cruse, R. M. (1994), Crop rotations for the 21st century, *Advances in Agronomy*, 53, 1–45.

Kassam, A., Friedrich, T., Derpsch, R. and Kienzle, J. (2015), Overview of the worldwide spread of conservation agriculture, *Field Actions Sci. Rep.: J. Field Actions*, 8, 1–10.

Kay, J. G. and Owen., H. (1973), Transmission of *Rhynchosporium secalis* on barley grain, *Trans. Br. Mycol. Soc.*, 60, 405–11.

Published by Burleigh Dodds Science Publishing Limited, 2019.

Kinane, J. and Lyngkaer, M. F. (2002), Effect of barley-legume intercrop on disease frequency in an organic farming system, *Proc. 6th Conf. EFPP 2002, Prague. Plant Prot. Sci.*, 38, 227–31.

Kosinski, S. M., King, J. R., Harker, K. N., Turkington, T. K. and Spaner, D. (2011), Barley and triticale underseeded with a kura clover living mulch: Effects on weed pressure, disease incidence, silage yield, and forage quality, *Can. J. Plant Sci.*, 91, 677–87.

Krupinsky, J. M., Bailey, K. L., McMullen, M. P., Gossen, B. D., Turkington, T. K. (2002), Managing plant disease risk in diversified cropping systems, *Agron. J.*, 94, 198–209.

Krupinsky, J. M., Tanaka, D. L., Lares, M. T. and Merrill, S. D. (2004), Leaf spot diseases of barley and spring wheat as influenced by preceding crops, *Agron. J.*, 96, 259–66.

Krupinsky, J. M., Tanaka, D. L., Merrill, S. D., Liebig, M. A. and Hanson, J. D. (2006), Crop sequence effects of 10 crops in the northern Great Plains, *Agric. Syst.*, 88, 227–54.

Krupinsky, J. M., Halvorson, A. D., Tanaka, D. L. and Merrill, S. D. (2007), Nitrogen and tillage effects on wheat leaf spot diseases in the Northern Great Plains, *Agron. J.*, 99, 562–9.

Kumar, K., Holtz, M. D., Xi, K. and Turkington, T. K. (2013), Overwintering potential of the stripe rust pathogen (*Puccinia striiformis*) in central Alberta, *Can. J. Plant Pathol.*, 35, 304–14.

Kutcher, H. R., Johnston, A. M., Bailey, K. L. and Malhi, S. S. (2011), Managing crop losses from plant diseases with foliar fungicides, rotation, and tillage on a Black Chernozem in Saskatchewan, Canada, *Field Crops Res.*, 124, 205–12.

Kutcher, H. R., Brandt, S. A., Smith, E. G., Ulrich, D., Malhi, S. S. and Johnston, A. M. (2013), Blackleg disease of canola mitigated by resistant cultivars and four-year crop rotations in western Canada, *Can. J. Plant Pathol.*, 35, 209–21.

Leduc, P. (Ed.). (1994), *Direct Seeding Manual*, Prairie Agricultural Machinery Institute and Saskatchewan Soil Conservation Association, Humboldt, Sask.

Lee, H. K., Tewari, J. P. and Turkington, T. K. (1999), Histopathology and isolation of *Rhynchosporium secalis* from infected barley seed, *Seed Sci. Tech.*, 27, 477–82.

Luke, H. H., Pfahler, P. L. and Barnett, R. D. (1983), Control of *Septoria nodorum* on wheat with crop rotation and seed treatment, *Plant Dis.*, 67, 949–51.

Lupwayi, N. Z., Rice, W. A. and Clayton, G. W. (1998), Soil microbial diversity and community structure under wheat as influenced by tillage and crop rotation, *Soil Biol. Biochem.*, 30, 1733–41.

Luz, W. C. da, Bergstrom, G. C. and Stockwell, C. A. (1998), Seed-applied bioprotectants for control of seedborne *Pyrenophora tritici-repentis* and agronomic enhancement of wheat, *Can. J. Plant Pathol.*, 20, 384–6.

Maloy, O. C. (2005), Plant disease management, *The Plant Health Instructor*, doi:10.1094/PHI-I-2005-0202-01, https://www.apsnet.org/edcenter/intropp/topics/Documents/PlantDiseaseManagement.aspx, (accessed 18 June 2018).

Manitoba Agriculture, Food and Rural Initiatives (MAFRI). (2008), Soil management guide, https://www.gov.mb.ca/agriculture/environment/soil-management/soil-management-guide/pubs/soil-management-guide.pdf (accessed 11 June 2018).

Mathre, D. E. (1997), *Compendium of Barley Diseases*, 2nd Edition, APS Press, St Paul, MN, 90pp.

Mayfield, A. H. and B. G. Clare. (1984a), Survival over summer of *Rhynchosporium secalis* in host debris in the field, *Aust. J. Agric. Res.*, 35, 789–97.

Mayfield, A. H. and B. G. Clare. (1984b), Effects of common stubble treatments and sowing sequences on scald disease (*Rhynchosporium secalis*) in barley crops, *Aust. J. Agric. Res.*, 35, 799–805.

Mazzilli, S. R., Ernst, O. R., de Mello, V. P. and Perez, C. A. (2016), Yield losses on wheat cops associated to the previous winter crop: Impact of agronomic practices based on on-farm analysis, *Eur. J. Agron.*, 75, 99–104.

McDonald, B. A. and Linde, C. (2002a), Pathogen population genetics, evolutionary potential, and durable resistance, *Annu. Rev. Phytopathol.*, 40, 349–79.

McDonald, B. A and Linde, C. (2002b), The population genetics of plant pathogens and breeding strategies for durable resistance, *Euphytica*, 124, 163–80.

Menzies, J. and Gilbert, J. (2003), Diseases of wheat, *in* Bailey, K. L., Gossen, B. D., Gugel, R. K. and Morrall, R. A. A. (Eds), *Diseases of Field Crops in Canada*, 3rd Edition, Canadian Phytopathological Society, Saskatoon, Saskatchewan, pp. 94–128.

Milus, E. A. and Chalkley, D. B. (1997), Effect of previous crop, seedborne inoculum, and fungicides on development of Stagonospora blotch, *Plant Dis.*, 81, 1279–83.

Mundt, C. C. (2002), Use of multiline cultivars and cultivar mixtures for disease management, *Annu. Rev. Phytopathol.*, 40, 381–410.

Nagarajan, S. and Singh, D. V. (1990), Long-distance dispersion of rust pathogens, *Annu. Rev. Phytopathol.* 28, 139–53.

Newton A. C. and Swanston, J. S. (1999), Cereal variety mixtures reducing inputs and improving yield and quality – why isn't everybody growing them? Scottish Crop Research Institute Annual Report for 1998.

Newton, A. C., Swanston, J. S., Guy, D. C. and Ellis, R. P. (1998), The effect of cultivar mixtures on malting quality in winter barley, *J. Inst. Brew.*, 104, 41–5.

Newton, A. C., Begg, G. S. and Swanston, J. S. (2009), Deployment of diversity for enhanced crop function, *Ann. Appl. Biol.*, 154, 309–22.

Paddock, W. C. (1967), Phytopathology in a hungry world, *Ann. Rev. Phytopathol.*, 5, 375–90.

Palti, J. (1981), *Cultural Practices and Infectious Crop Diseases*, Springer, Berlin, 243pp.

Parker, E. C. (1915), *Field Management and Crop Rotation*, Webb Publishing Co., St Paul, MN.

Parry, D. W., Jenkinson, P. and McLeod, L. (1995), *Fusarium* ear blight (scab) in small grain cereals – a review, *Plant Pathology*, 44, 207–38.

Pedersen, E. A. and Hughes, G. R. (1992), The effect of crop rotation on development of the septoria disease complex on spring wheat in Saskatchewan, *Can. J. Plant Pathol.*, 14, 152–8.

Pittelkow, C. M., Liang, X., Linquist, B. A., Van Groenigen, K. J., Lee, J., Lundy, M. E., Van Gestel, N., Six, J., Venterea, R. T. and van Kessel, C. (2015), Productivity limits and potentials of the principles of conservation agriculture, *Nature*, 517, 365–8.

Pridham, J. C. and Entz, M. H. (2008), Intercropping spring wheat with cereal grains, legumes and oilseeds fails to improve productivity under organic management, *Agron. J.*, 100, 1436–42.

Priestly, R. H. (1981), Choice and deployment of resistant cultivars for cereal disease control, *in* J. F. Jenkyn and R. T. Plumb (Ed.), *Strategies for the Control of Cereal Disease*, Blackwell Science Publishing, Oxford, UK.

Priestly, R. H. and Wolfe, M. S. (1977), Crop protection by cultivar diversification, *in Proceedings 1977 British Crop Protection Conference – Pests and Diseases*, Volume 1, pp. 135–40.

Reed, H. E. (1957), Studies on barley scald, Bull. Univ. Tenn. Agric. Exp. Stn., No. 268.

Rees, R. G., Platz, G. J. and Mayer, R. J. (1982), Yield losses in wheat from yellow spot: Comparison of estimates derived from single tillers and plots, *Aust. J. Agric. Res.*, 33, 899–908.

REX Consortium (2016), Combining selective pressures to enhance the durability of disease resistance genes, *Front. Plant Sci.*, 7, 1916.

Rothrock, C. S. (1992), Tillage systems and plant disease, *Soil Sci.*, 154, 308–15.

Sadler, E. J. and Turner, N. C. (1994), Water relationships in a sustainable agriculture system, *in* J. L. Hatfield and D. L. Karlen (Eds), *Sustainable Agriculture Systems*, CRC Press, Inc., Boca Raton, FL, pp. 21–46.

Sawinska, Z., Malecka, I. and Blecharczyk, A. (2006), Impact of previous crops and tillage systems on health status of winter wheat, *Electron. J. Pol. Agric. Univ.*, 9, #51, http://www.ejpau.media.pl/volume9/issue4/art-51.html (accessed 11 June 2018).

Schilder, A. M. C. and Bergstrom, G. C. (1995), Seed transmission of *Pyrenophora tritici-repentis*, causal fungus of tan spot of wheat, *Eur. J. Plant Pathol.*, 101, 81–91.

Schuh, W. (1990), The influence of tillage systems on incidence and spatial pattern of tan spot of wheat, *Phytopathology*, 80, 804–7.

Shah, D., Bergstrom, G. C. and Ueng, P. P. (1995), Initiation of Septoria nodorum blotch epidemics in winter wheat by seedborne *Stagonospora nodorum*, *Phytopathology*, 85, 452–7.

Sheridan, J. E., Grbavac, N. and Ballard, L. (1983), Strategies for controlling net blotch of barley, *Proc. 36th N. Z. Weed and Pest Control Conf.*, pp. 242–6.

Shipton, W. A., Khan, T. N. and Boyd, W. J. R. (1973), Net blotch of barley, *Rev. Plant Pathol.*, 52, 269–90.

Singh, M. and Chand, J. N. (1985), Studies on the survival of *Helminthosporium teres* incitant of net blotch of barley, *Indian Phytopath.*, 38, 659–61.

Skoropad, W. P. (1959), Seed and seedling infection of barley by *Rhynchosporium secalis*, *Phytopathology*, 49, 623–6.

Smith, E. G., Kutcher, H. R., Brandt, S. A., Ulrich, D., Malhi, S. S. and Johnston, A. M. (2013), The profitability of short-duration canola and pea rotations in western Canada, *Can. J. Plant Sci.*, 93, 933–40.

Smith, E. G., Turkington, T. K., O'Donovan, J. T., Edney, J. T., Juskiw, P. E., McKenzie, R. H., Clayton, G. W., Xi, K., May, W. E., Irvine, R. B., Brandt, S., Johnson, E. N. and Perkovic, S. (2015), Influence of production systems on return and risk from malting barley production in Western Canada, *Can. J. Plant Sci.*, 96, 339–46.

Soane, B. D., Ball, B. C., Arvidsson, J., Basch, G., Moreno, F. and Roger-Estrade, J. (2012), No-till in northern, western and south-western Europe: A review of problems and opportunities for crop production and the environment, *Soil Till. Res.*, 118, 66–87.

Stack, R. W. (2000), Return of an old problem: Fusarium head blight of small grains, *APSnet Plant Health Reviews*, http://www.apsnet.org/publications/apsnetfeatures/Pages/headblight.aspx, (accessed 15 August 2018).

Steiner, J. L. (1994), Crop residue effects on water conservation, *in* P. W. Unger (Ed.), *Managing Agricultural Residues*, CRC Press, Inc., Boca Raton, FL, pp. 41–76.

Sullivan, P. (2001), Intercropping principles and practices: Agronomy systems guide, Appropriate technology transfer for rural areas, National Center for Appropriate Technology, Fayetteville, AR, http://www.attra.org/attra-pub/intercrop.html (accessed 18 June 2018).

Sutton, J. C. (1982), Epidemiology of wheat head blight and maize ear rot caused by *Fusarium graminearum*, *Can. J. Plant Pathol.*, 4, 195–209.

Tanaka, D. L., Krupinsky, J. M., Liebig, M. A., Merrill, S. D., Ries, R. E., Hendrickson, J. R., Johnson, H. A. and Hanson, J. D. (2002), Dynamic cropping systems, *Agron. J.*, 94, 957–61.

Tekauz, A. (2003), Diseases of barley, *in* K. L. Bailey, B. D. Gossen, R. K. Gugel and R. A. A. Morrall (Ed.), *Diseases of Field Crops in Canada*, 3rd Edition, Canadian Phytopathological Society, Saskatoon, Saskatchewan, pp. 30–53.

Turkington, T. K. (2003), Diseases of field crops in Canada, *in* K. L. Bailey, B. D. Gossen, R. K. Gugel, and R. A. A. Morrall (Ed.), *Diseases of Field Crops in Canada*, 3rd Edition, Canadian Phytopathological Society, Saskatoon, Saskatchewan, pp. 9–14.

Turkington, T. K., Clear, R. M., Burnett, P. A., Patrick, S. K., Orr, D. D. and Xi, K. (2002), Fungal plant pathogens infecting barley and wheat seed from Alberta, 1995–1997, *Can. J. Plant Pathol.*, 24, 302–8.

Turkington, T. K., Xi, K., Tewari, J. P., Lee, H. K., Clayton, G. W. and Harker, K. N. (2005), Cultivar rotation as a strategy to reduce leaf diseases under barley monoculture, *Can. J. Plant Pathol.*, 27, 283–90.

Turkington, T. K., Xi, K., Clayton, G. W., Burnett, P. A., Klein-Gebbinck, H., Lupwayi, N. Z., Harker, K. N. and O'Donovan, J. T. (2006), Impact of crop management on leaf diseases in Alberta barley fields, 1995–1997, *Can. J. Plant Pathol.*, 28, 441–9.

Turkington, T. K., O'Donovan, J. T., Edney, M. J., Juskiw, P. E., McKenzie, R. H., Harker, K. N., Clayton, G. W., Xi, K., Lafond, G. P., Irvine, R. B., Brandt, S., Johnson, E. N., May, W. E. and Smith, E. (2012), Effect of crop residue, nitrogen rate and fungicide application on malting barley productivity, quality, and foliar disease severity, *Can. J. Plant Sci.*, 92, 577–88.

Turkington, T. K., Xi, K., Harker, K. N., O'Donovan, J. T., Blackshaw, R., McAllister, T. and Lupwayi, N. Z. (2015a), The impact of barley variety rotation, mixtures, and intercropping on leaf disease and silage production, *Can. J. Plant Pathol.*, 38, 131–2 (Abstr.).

Turkington, T. K., Xi, K., Harker, K. N., O'Donovan, J. T., Blackshaw, R., McAllister, T. and Lupwayi, N. Z. (2015b), The impact of barley variety rotation, mixtures, and intercropping on leaf disease and silage production, *Proceedings of the XVIII International Plant Protection Congress (IPPC) 2015*, 24–27 August 2015, Free University Berlin, Henry Ford Building, Berlin, Germany, ID: O NOC I-1 (oral presentation).

Vilich-Meller, V. (1992), Mixed cropping of cereals to suppress plant diseases and omit pesticide applications, *Bio. Agri. Hort.*, 8, 299–308.

Watkins, J. E. and Boosalis, M. G. (1994), Plant disease incidence as influenced by conservation tillage systems, *in* P. W. Unger (Ed.), *Managing Agricultural Residues*, CRC Press, Inc., Boca Raton, FL, pp. 261–83.

Wiese, M. V. (1987), *Compendium of Wheat Diseases*, 2nd Edition, APS Press, St Paul, MN, 112pp.

Wilkins, D. E., Klepper, B. L. and Rasmussen, P. E. (1988), Management of grain stubble for conservation-tillage systems, *Soil Till. Res.*, 12, 25–35.

Wolfe, M. S. (1978), Some practical implications of the use of cereal variety mixtures, *in* P. R. Scott and A. Bainbridge (Eds), *Plant Disease Epidemiology*, Blackwell Science Publishing, Oxford, UK, pp. 201–7.

Wolfe, M. S. (1985), The current status and prospects of multiline cultivars and variety mixtures for disease resistance, *Annu. Rev. Phytopathol.*, 23, 251–73.

Wolfe, M. S. (2000), Crop strength through diversity, *Nature*, 406, 681–2.

Wolfe, M. S. (2002), The role of functional biodiversity in managing pest and diseases in organic production systems, *in the BCPC Conference, Pest and Diseases 2002*, Volume 1, pp. 531–616, Proceedings of the British Crop Protection Council, Brighton, UK, 18–21 November 2002, The British Crop Protection Council, Surrey, UK.

Xi, K. and Burnett, P. A. (1997), Staining paraffin-embedded sections of scald of barley before paraffin removal, *Biotech. Histochem.*, 72, 173–7.

Zadoks, J. C. and Schein, R. D. (1979), *Epidemiology and Plant Disease Management*, Oxford University Press, New York.

Zhan, J., Thrall, P. H., Papaïx, J., Xie, L. and Burdon, J. J. (2015), Playing on a pathogen's weakness: Using evolution to guide sustainable plant disease control strategies, *Annu. Rev. Phytopathol.*, 53, 19–43.

Zhu, J-H., Dong, Y., Xiao, J-X., Zheng, Y. and Tang, L. (2017), Effects of N application on wheat powdery mildew occurrence, nitrogen accumulation and allocation in intercropping system, *Chinese J. Applied Ecol.*, 28, 3985–93. In Chinese with English abstract.

Published by Burleigh Dodds Science Publishing Limited, 2019.

Integrated wheat disease management

Stephen N. Wegulo, University of Nebraska-Lincoln, USA

1 Introduction

Wheat is grown on the world's largest crop area. It is the most widely consumed grain and most internationally traded agricultural commodity (Fischer, 2009). Diseases are a major constraint in wheat production. In 2001–3, worldwide potential losses in wheat due to diseases were estimated at 18.1% and actual losses were estimated at 12.6% (Oerke, 2006). Four groups of microorganisms cause biotic diseases: bacteria, fungi, nematodes and viruses. These microorganisms reside in the soil, crop residues or living plant hosts. All parts of the wheat plant (roots, crown, stem, leaves and spike) can be affected by disease.

Because of the huge economic losses caused by diseases in wheat production, it is essential to develop and implement management tactics that prevent or reduce yield loss. Integrated disease management combines two or more tactics to manage plant diseases. It is based on the concept of integrated pest management, which is defined as '. . . a sustainable approach to managing pests by combining biological, cultural, physical, and chemical tools in a way that minimizes economic, health, and environmental risks' (Anonymous, 1994). These tools or tactics include scouting, disease identification, variety selection, cultural practices, chemical control and biological control (El Khoury and Makkouk, 2010; Hollier and Hershman, 2008; Krupinsky et al., 2002). In addition, the use of disease forecasting systems can enable growers to deploy management tactics in a timely manner and only when necessary. This

http://dx.doi.org/10.19103/AS.2016.0004.23
© Burleigh Dodds Science Publishing Limited, 2017. All rights reserved.

chapter reviews these tactics individually using selected research examples and emphasizes the integration of two or more tactics to more effectively manage wheat diseases.

2 Scouting and disease identification

Scouting for early disease presence is a critical step in the successful implementation of an integrated disease management programme. Wheat fields should be scouted regularly starting early in the growing season. The frequency of scouting depends on weather conditions; it is necessary to scout more frequently if disease-favourable weather is forecast. To get an accurate assessment of the diseases present and their levels, a representative area of the field is scouted. This is achieved by using one of several patterns to walk through the field. Some patterns that are frequently used are X, W and Z (Hollier and Hershman, 2008). During the process, frequent stops are made and plants are examined for disease symptoms in the upper and lower canopies of the wheat crop. If chlorosis and stunting are observed, a few plants should be pulled and the roots examined. Therefore, it is essential that the scout has the knowledge or training necessary to identify diseases based on symptoms.

If symptoms are present but the disease cannot be identified, samples are collected and submitted to a professional laboratory or clinic for diagnosis. The time required for diagnosis can be shortened by using electronic devices that take and transmit digital images of symptoms. These digital images can be sent electronically to professionals who can provide a diagnosis quickly if the symptoms are unique to a particular disease. A correct diagnosis ensures the correct disease management tactic is applied. Often a laboratory examination or procedure is necessary to arrive at a correct diagnosis. The scout also notes parts of the field that may be more heavily affected by specific diseases or that may have a disease of interest or concern. To this end, it is useful for the scout to have a GPS unit to record coordinates of places in the field that may need special attention. Information gained from scouting is used to plan if and when a disease management measure such as fungicide application can be implemented. This information can also be used to build a field history that can be used in making future disease management decisions (Hollier and Hershman, 2008; Krupinsky et al., 2002).

3 Variety selection

Farmers select wheat varieties based on multiple factors. In Kansas, Barkley and Porter (1996) used regression analysis to show that the farmers considered the following factors when selecting wheat varieties during the period from 1974 to 1993: end-use quality, production characteristics (yield, yield stability and resistance to diseases, pests and adverse environmental conditions) and past production decisions. In Denmark, Detlefsen and Jensen (2004) noted that based on experience from the extension service, farmers select winter wheat varieties based mainly on yield, and this may be suboptimal because other characteristics are neglected. They described a stochastic optimization model for Danish farmers to use when selecting winter wheat varieties. The model requires data on yield, susceptibility to diseases and various quality characteristics and calculates, for

© Burleigh Dodds Science Publishing Limited, 2017. All rights reserved.

each winter wheat variety, the expected net revenue by subtracting disease treatment and fertilizer application costs from the expected gross revenue. In the UK, the National Institute of Agricultural Botany (NIAB) interactive cereal variety Gross Margin Model (Nelson and Meikle, 2001) was developed to calculate the gross margin for individual varieties of winter wheat, winter barley and spring barley by subtracting variable production costs (seed, fertilizer, fungicide and other sprays) from output (yield × grain price + area payments). In the model, the fungicide regime for each variety is tailored after consideration of the variety's disease resistance level and instruction on diseases of major concern.

These examples of variety selection tools and criteria demonstrate the importance of genetic resistance, which is one of the most cost-effective strategies for managing wheat diseases. Commercially available varieties can differ widely in their levels of resistance to a particular disease and for some diseases, little or no varietal resistance may exist. It is best to select varieties that differ in genetic background. Because of genetic differences, wheat varieties will react differently to diseases and some varieties will mature sooner or later than others. Planting several varieties with different genetic backgrounds, each variety in a separate field, is a strategy that can reduce losses due to diseases. For example, if only one variety is planted and it happens to be susceptible to a predominant disease during the growing season, yield loss can be much greater than if two or several varieties that have different levels of resistance were planted. Fusarium head blight (FHB), caused by *Fusarium graminearum* and other species in the genus *Fusarium* and related genera, is a disease that affects wheat spikes. Because there is only a short window (flowering) during which the infection of wheat by *F. graminearum* occurs, if two or several varieties differing in flowering dates are planted, there is an increased probability that one or a few of the varieties will escape heavy infections. Planting varieties that differ in maturity dates can also facilitate timely harvesting of each variety. A similar but less commonly used strategy consists of planting, in the same field, varieties that differ in agronomic characters and disease resistance levels but are sufficiently similar to be grown together. Such plantings, known as variety mixtures, have been demonstrated to reduce the intensity of diseases in small grain crops (Mundt, 2002; Wolfe, 1985).

4 Cultural practices

Cultural control of plant diseases is preventive in nature and is achieved by adjusting crop management practices to prevent or minimize disease development (Anonymous, 1968). In wheat production, the most common cultural practices used to manage diseases include crop rotation, residue management, tillage, planting date, nutrient management and irrigation management.

4.1 Crop rotation

Crop rotation is the successive planting of different crops in the same field. It is among the oldest and most widely used cultural practices in agricultural production and has been utilized to a great extent in controlling residue- and soil-borne pathogens of cereal crops (Ogle and Dale, 1997). The benefits of break crops in wheat production (reviewed by Angus et al., 2015; Kirkegaard et al., 2008) include disease control and increased residual water and nutrients available to the wheat crop planted following the break crop. Average

© Burleigh Dodds Science Publishing Limited, 2017. All rights reserved.

yield benefits of up to 20% or more from break crops have been reported, and grain legumes have been found to provide the greatest benefits (Angus et al., 2015; Kirkegaard et al., 2008).

The effects of crop rotation as a disease management tactic are illustrated using take-all of wheat caused by the fungus *Gaeumannomyces graminis* var. *tritici* (*Ggt*). *Ggt* survives as mycelium in wheat residue (Bockus and Shroyer, 1998; Paulitz, 2010). Roots are infected by mycelium growing from the residue. A brown to black rot is initiated and develops upwards into the crown and up to the culm base. Infected plants are stunted and mildly chlorotic with fewer tillers. Some tillers die prematurely, resulting in distinctively bleached and sterile spikes with shrivelled kernels. Infected plants pull up easily or break near the soil line due to brittle stem bases. Take-all can cause up to 20% yield loss (Murray et al., 1998). The effect of crop rotation on take-all has been studied extensively. This is because host plant resistance and chemical control have not been effective in controlling the disease, whereas crop rotation has consistently reduced disease intensity and yield loss (Cook, 2003).

In a crop rotation study that surveyed 87 commercial fields and 10 farms in New Zealand between 2003 and 2009 (van Toor et al., 2013), *Ggt* soil concentrations and take-all severity were low in the first wheat crop following a non-host crop (oat, ryegrass or clover) but increased thereafter, reaching a maximum in the second consecutive wheat crop. Double breaks consisting of two consecutive non-host crops resulted in much lower *Ggt* concentrations and take-all severity than single breaks. In France, Ennaïfar et al. (2005) found that when combined with conventional tillage, the summer fallow non-host crops oat, mustard and ryegrass reduced take-all intensity in the following wheat crop compared to volunteer wheat. In a three-year crop rotation study in Germany, Sieling et al. (2007) grew winter wheat following oilseed rape or winter wheat. Results showed that oilseed rape as a preceding crop significantly reduced take-all and resulted in 13% more yield compared to winter wheat. In an irrigated cropping system in the Pacific northwest of the U.S., a three-year crop rotation of winter wheat-spring barley-canola under no-till was compared to continuous annual winter wheat planted after stubble burning and mouldboard ploughing (Paulitz et al., 2010). Take-all severity in the third year was two to three units higher in continuous winter wheat compared to the diversified no-till rotation treatments. In the southern U.S., a long-term crop rotation study showed that take-all severity was high in rotations with continuous wheat, but was significantly reduced when one crop of winter canola preceded winter wheat (Cunfer et al., 2006). These take-all examples illustrate the effectiveness of crop rotation in controlling a residue-borne wheat disease. The effectiveness of crop rotation in controlling residue- and soil-borne diseases of wheat has been demonstrated for other diseases including FHB (Dill-Macky and Jones, 2000), tan spot (*Pyrenophora tritici-repentis*) (Bockus and Classen, 1992) and *Cephalosporium* stripe (*Cephalosporium gramineum*) (Latin et al., 1982).

4.2 Tillage and residue management

Conventional or clean tillage buries crop residues, which reduces the incidence and severity of residue- and soil-borne diseases. This is because conventional tillage breaks up and buries crop residues on which plant pathogens survive, allowing native soil microorganisms to rapidly decay the residue, resulting in pathogen death (Bockus and Shroyer, 1998; Watkins and Boosalis, 1994). Additional advantages of conventional tillage include reduction in surface compaction, a warmer seedbed in the spring, breaking up of

© Burleigh Dodds Science Publishing Limited, 2017. All rights reserved.

hardpans and incorporation of pesticides and fertilizers (Krupisnky et al., 2002). However, this type of tillage causes soil erosion and does not conserve soil moisture. Because of this and government-mandated soil conservation programmes in some countries such as the U.S. (Simmons and Nafziger, 2009), many farmers practice conservation tillage to conserve soil and moisture, especially in regions or areas where soil moisture is limiting. In conservation tillage, at least 30% of the soil surface is covered with residues where soil erosion by water is dominant or the soil surface is covered with at least 1 metric ton of residues per hectare where soil erosion by wind is dominant (Anonymous, 1990). Conservation tillage consists of several different tillage systems. They are no-till or zero tillage and various types of reduced tillage including strip-till, ridge-till and mulch-till (Simmons and Nafziger, 2009; Unger, 1994). In addition to soil and moisture conservation, conservation tillage reduces costs because less fuel is used during farming operations and improves soil structure by adding organic matter and positively influencing soil physical properties such as porosity, aggregate stability and bulk density. Soil water factors such as infiltration, retention and conductivity are also positively influenced. The residues provide nutrients to macro- and microorganisms which improve soil physical conditions by converting residues and organic matter into stable materials (Bailey and Duczek, 1996; Bockus and Shroyer, 1998; Unger, 1994).

Examples of residue-borne diseases of wheat are tan spot, *Septoria tritici* blotch, *Stagonospora nodorum* blotch, take-all, *Cephalosporium* stripe and FHB (Bockus, 1998; Bockus and Shroyer, 1998; Dill-Macky, 2010). The pathogens that cause these diseases survive on crop residues between harvest and the next growing season. Tillage and residue management practices can reduce or increase incidence and severity depending on the disease. The effects of these management practices on four wheat diseases (*Cephalosporium* stripe, tan spot, FHB and take-all) are reviewed.

4.2.1 *Cephalosporium* stripe

Cephalosporium stripe is a vascular disease of wheat. The causal fungus, *C. gramineum*, survives intercrop periods as mycelium in wheat straw (Bockus and Shroyer, 1998; Mundt, 2010). Spores are formed in the fall and winter and passively move through the soil to the root zone where they infect the roots. Subsequently, the pathogen enters the vascular system and plugs it. Conspicuous longitudinal and chlorotic stripes form on the leaves during jointing and heading of the wheat crop. The disease can cause up to 80% yield loss (Mundt, 2010). In a field in which continuous winter wheat was cultivated in Michigan, Wiese and Ravenscroft (1975) found that when residue from the previous year's crop was thoroughly removed before planting, both *C. gramineum* and *Cephalosporium* stripe (Cs) were nearly eliminated. Residue disked into the soil before planting supported four times the number of *C. gramineum* propagules and twice the incidence of Cs compared to deep ploughing. In Kansas, Bockus et al. (1983) compared the effects of five residue management practices on Cs during three years of continuous winter wheat cropping. These practices were burn and disk, plough, disk, chop and disk and direct drill. Results showed that burn and disk was the most effective practice in reducing Cs, with a three-year average of 12.8% Cs incidence followed by plough (24.2%), disk (29.6%), chop and disk (36.7%) and direct drill (46%). It is clear from these two studies (Wiese and Ravenscroft, 1975; Bockus et al., 1983) that tillage and residue management practices that reduce or eliminate residue also reduce Cs, whereas those that leave crop residue on the soil surface or in the topsoil favour pathogen multiplication and disease development.

© Burleigh Dodds Science Publishing Limited, 2017. All rights reserved.

4.2.2 Tan spot

Another residue-borne disease whose intensity can be influenced by tillage and residue management practices is tan spot caused by *P. tritici-repentis*. The pathogen survives as mycelium in wheat straw. It forms sexual fruiting structures known as pseudothecia on the straw during the summer and fall. Pseudothecia mature by the spring and release ascospores which cause primary infections on the foliage. Conidia produced in the primary lesions are spread by wind and cause infections over large areas in the field (Dill-Macky, 2010; Bockus and Shroyer, 1998). Yield losses of up to 49% have been reported to be caused by tan spot (Rees et al., 1982; Wegulo et al., 2009). Because the pathogen survives on wheat residue, tillage practices that leave crop residue on the soil surface promote intercrop survival of the pathogen, resulting in severe epidemics (Bockus and Shroyer, 1998). In Kansas, Bockus and Claassen (1992) compared tan spot severity among three residue management practices (mouldboard plough, chisel plough and no-till) in continuous winter wheat. They found that there was significant and consistent reduction of tan spot severity in the mouldboard plough treatment compared to the chisel and no-till treatments, indicating that tan spot was more severe when residue was left on the soil surface. In Australia, Summerell et al. (1988) had earlier shown that the severity of tan spot was greatly increased when wheat residue was retained on the soil surface, whereas the incorporation of residue slowed disease development and burning the residue was most effective in controlling the disease. In another study conducted by Summerell and Burgess (1989) in which wheat residue was retained on the soil surface, incorporated or buried, it was shown that burying was detrimental to the survival of *P. tritici-repentis*, whereas it survived longest on residue retained on the soil surface.

4.2.3 Fusarium head blight

FHB affects the wheat spike. The causal pathogens, of which the major one is *F. graminearum*, survive and overwinter as mycelium or chlamydospores in crop residue. Corn, wheat and sorghum are highly suitable residues for the survival of FHB-causing fungi. Perithecia mature during the spring and release ascospores which infect spikes during anthesis if favourable environmental conditions (wet weather before and during anthesis) prevail. Infected spikes are sterile or contain shrivelled, chalky white or pink kernels known as *Fusarium*-damaged kernels (FDK). The pathogens produce mycotoxins which contaminate grain. The most commonly produced mycotoxin is deoxynivalenol (DON). Management strategies and tactics for FHB and DON, including tillage and residue management, were recently reviewed by Wegulo (2012) and Wegulo et al. (2015). In Canada, Khonga and Sutton (1988) found that when *F. graminearum*-infested corn ears, corn stems and wheat spikes, stems and grain were placed above, on or below the soil surface, perithecia were produced on residue above the soil for up to three years, but were not produced on buried residue. In Minnesota, Dill-Macky and Jones (2000) showed that FHB incidence and severity were lower in mouldboard-ploughed plots compared to no-till or chisel-ploughed plots. Maiorano et al. (2008) assessed the influence of corn residues and tillage on *Fusarium* infection and DON production in winter wheat in Italy. There was a strong correlation between residue amounts and DON in grain. DON concentrations were higher in no-till fields and in residues retained on the soil surface than in tilled fields and buried residues.

© Burleigh Dodds Science Publishing Limited, 2017. All rights reserved.

4.2.4 Take-all

Take-all is favoured by reduced tillage because infested wheat residue in which *Ggt* survives lasts longer under reduced tillage compared to conventional tillage (Bockus and Shroyer, 1998). However, residue, in addition to the direct influence on take-all by providing a habitat for the pathogen, can indirectly influence disease intensity. In Kansas, Bockus et al. (1994) incorporated artificially prepared inoculum of *Ggt* 10 cm deep in field plots one month after wheat harvest and two months before planting wheat and either shaded the plots with wheat straw or left them bare. Take-all severity and yield loss were greater in shaded than in bare plots and this was attributed to a lower soil temperature in shaded plots which prevented the thermal inactivation of *Ggt*, indicating that management practices such as zero tillage that increase shading tend to promote inoculum survival and disease development. Some studies have demonstrated the opposite effect of tillage on take-all. In Canada, Bailey et al. (1992) found the incidence of take-all in winter wheat to be higher under conventional tillage compared to minimum or zero tillage. They suggested that soil disturbance by cultivation can alter the rhizosphere microclimate and reduce the establishment of beneficial microorganisms while allowing take-all to colonize. In the same study, common root rot caused by *Bipolaris sorokiniana* was reduced by minimum and zero tillage. In New Zealand, Cromey et al. (2008) found take-all incidence to be higher in disked than in no-till plots. They hypothesized that disking may have spread take-all inoculum in the soil, leading to a higher incidence of the disease than in no-till plots, or greater aeration and exposure to light by disking could have resulted in a longer period of saprophytic survival of *Ggt*.

4.3 Planting date

Adjustment of planting date is a tactic commonly used to manage certain wheat diseases. Its effectiveness is related to the prevailing environmental conditions during and after seeding, mainly temperature. In winter wheat, delaying the sowing date in autumn can reduce the damage caused by the following root, crown and culm rot diseases: common root rot, *Fusarium* crown rot (*F. culmorum*, *F. pseudograminearum*), take-all (*Ggt*), *Cephalosporium* stripe and eyespot (*Oculimacula yallundae*, *Oculimacula acuformis*) (Smiley et al., 2009). Damage caused by the virus diseases such as wheat streak mosaic, *Triticum* mosaic and barley yellow dwarf (BYD) can also be reduced by delaying autumn sowing (Gray, 2010; Hunger, 2010; Plumb and Johnstone, 1995). Early seeding can reduce damage caused by common bunt (*Tilletia caries*, *Tilletia laevis*) (Carris, 2010). Specific examples of research on the effects of planting date on wheat diseases are provided for root and crown rot fungal diseases, above-ground fungal diseases and virus diseases.

4.3.1 Root and crown rot fungal diseases

Date of planting studies conducted in western Nebraska over a 15-year period (Fenster et al., 1972) showed that the severity of root and crown rot of winter wheat caused by *B. sorokiniana*, *F. graminearum*, or both pathogens decreased as planting date was delayed. At one representative location in black fallow plots, disease severity assessed in April on roots and in crowns was high in wheat planted on 20 August, medium in wheat planted on 1 September and low (crowns) and medium (roots) in wheat planted on 10 September. Crowns had no disease symptoms in wheat planted on 20 and 30 September.

© Burleigh Dodds Science Publishing Limited, 2017. All rights reserved.

In Oregon, Smiley (2009) planted winter wheat on sequential dates from early September to December and measured the incidence of *Cephalosporium* stripe, *Fusarium* crown rot, eyespot, *Rhizoctonia* root rot and take-all during four consecutive cropping seasons. Incidence of all but *Rhizoctonia* root rot decreased as planting date was delayed. Incidence of *Fusarium* crown rot and eyespot was positively correlated with air and soil temperatures, whereas the incidence of take-all was positively correlated with soil temperature, indicating that planting later in the autumn at cooler temperatures reduced disease. Neither planting date nor weather variables had an effect on *Rhizoctonia* root rot. In Texas, common root rot indices were higher in winter wheat planted in September compared to October (Piccinni et al., 2001). In Idaho, Goates and Mercier (2011) evaluated the efficacy of the biocontrol agent *Muscodor albus* applied as a seed treatment in controlling common bunt in spring wheat seeded at two dates. The first seeding was done at the normal planting date when soil temperatures were cool and optimal for infection and disease development, and the second seeding was done three weeks later when soil temperatures were warmer and less favourable for infection and disease development. In the earlier planted wheat, incidence of diseased spikes was 44% in controls and 12% in seed treatments compared to 6% and 0%, respectively, in the later planted wheat. These results demonstrated that late sowing of spring wheat (equivalent of early sowing of winter wheat) when soil temperatures are warm can reduce damage caused by common bunt.

4.3.2 Above-ground fungal diseases

The diseases in the preceding examples of the effects of planting date on disease intensity are caused by pathogens that infect the roots or crowns of wheat plants. Adjustment of planting date in spring wheat can also influence the amount of above-ground diseases of wheat, especially those whose causal agents survive on crop residues. In Canada, Subedi et al. (2007) evaluated the effect of planting date of a commercial spring wheat cultivar on the severity of the residue-borne leaf spot disease complex consisting of tan spot, *S. tritici* blotch and *S. nodorum* blotch as well as on the incidence and severity of FHB. The wheat was planted on three dates from as early as 18 April to as late as 6 June with 10–17 days between the first and second planting dates and 10–21 days between the second and third planting dates. In four out of five site-years, the severity of leaf spot diseases on the penultimate leaf (leaf below the flag leaf) was significantly higher in one or both of the latter planting dates compared to the first planting date. This was also the case for the flag leaf in three out of five site-years. FHB intensity (incidence, severity and index) was higher as planting date was delayed beyond 9 May. Regression of FHB incidence on planting date clearly showed that FHB incidence increased with later planting. The increase in the intensity of leaf spot diseases and FHB with later planting was attributed to increasing temperatures and relative humidity as the growing season progressed. Wheat planted on the second date reached the 50% heading growth stage 5–15 days later than wheat planted on the first date. This growth stage was delayed by 8–19 days in wheat planted on the third date compared to wheat planted on the second date. Therefore, compared to the earliest planting, infections in later plantings occurred at warmer temperatures which, coupled with moisture, favour increased spore release from fruiting structures on crop residues as well as primary and secondary infections and disease development (Agrios, 2005; Shaner, 1981). In south Asia, Gurung et al. (2012) similarly showed that the severity of spot blotch and tan spot in four genotypes of wheat was higher in wheat sown 15 and 30 days later than the optimum planting date.

© Burleigh Dodds Science Publishing Limited, 2017. All rights reserved.

4.3.3 Virus diseases

Planting date can also influence the intensity of virus diseases of wheat. In South Dakota, Hesler et al. (2005) showed that aphid infestations and the resulting BYD were reduced in winter wheat planted on 20 September or later compared to wheat planted earlier. In one site-year, BYD incidence was reduced by 54% in wheat planted on 20 September compared to wheat planted on 31 August and by 40% compared to wheat planted on 10 September. In another site-year, BYD incidence was reduced by 42% in wheat planted on 27 September compared to wheat planted on 9 September and by 29% compared to wheat planted on 18 September. To demonstrate the effect of planting date on wheat streak mosaic in Nebraska (Staples and Allington, 1956), nine strips of winter wheat were planted perpendicular to a field of wheat curl mite (*Aceria tosichella*)-infested and diseased volunteer wheat. Each strip was planted on a different date starting from 17 August to 10 October. One hundred plants from each strip were collected on 1 December, and the disease incidence was determined by the manual inoculation of healthy wheat seedlings in the greenhouse. Disease incidence decreased with later planting dates, with 100% incidence in strips planted on 17 and 22 August, 24% incidence in the strip planted on the second to last date (3 October) and 6% incidence in the strip planted on the latest date (10 October). The reduction in disease incidence with later planting was attributed to decreased exposure of late planted wheat to viruliferous wheat curl mites which transmit wheat streak mosaic virus (WSMV) to healthy wheat plants.

4.4 Nutrient management

Plants require nutrients for growth and proper functioning of metabolic processes. Sixteen essential plant nutrients are broadly recognized. The macronutrients, required in large quantities, consist of hydrogen, carbon, oxygen, nitrogen, potassium, calcium, magnesium, phosphorus and sulphur. The micronutrients, required in smaller amounts, are chlorine, boron, iron, manganese, zinc, copper and molybdenum (Daroub and Snyder, 2007; Rice, 2007). Insufficient or excessive nutrients can increase the susceptibility of plants to diseases (Agrios, 2005; Graham, 1983; Huber, 1991). In wheat, for example, excessive nitrogen increases susceptibility to powdery mildew (Chen et al., 2007) and stripe rust (Devadas et al., 2015) and phosphorus deficiency increases susceptibility to *Pythium* root rot (Vanterpool, 1935). Therefore, balanced and adequate fertilization which optimizes growth and yield potential and enhances the crop's ability to resist pathogen attack is essential for successful wheat production. To achieve balanced and adequate nutrition provided by both macronutrients and micronutrients, chemical analysis of soil to determine which nutrients are needed in what amounts and the proper timing of fertilizer application are necessary. Both the amount and type of fertilizer applied can influence disease development (Huber, 1991; Huber et al., 1968; Huber and Watson, 1974). Fertilizers and organic amendments have been used to control certain diseases of wheat. The effects of various mineral nutrients on wheat diseases are detailed in Datnoff et al. (2007). Some of these effects are illustrated by the following examples.

In Washington State, Smiley and Cook demonstrated a reduction in severity of take-all in winter wheat plots treated with the ammonium form of nitrogen (NH_4^+-N) and an increase in severity of the disease in plots treated with nitrate nitrogen (NO_3^--N), confirming previous

© Burleigh Dodds Science Publishing Limited, 2017. All rights reserved.

results obtained in Idaho by Huber et al. (1968). Reis et al. (1982), also in Washington State, demonstrated a reduction in take-all in winter wheat when phosphorus, potassium, magnesium, zinc, copper, manganese and iron were made available to the roots in growth chamber and field studies. In Oregon, Taylor et al. (1983) showed that losses in winter wheat due to take-all were reduced when fertilizers containing phosphorus, chloride and NH$_4^+$-N were applied. In Canada, Subedi et al. (2007) found that spring wheat grown without nitrogen fertilization had a higher incidence of FHB and leaf spot diseases (tan spot, *S. tritici* blotch and *S. nodorum* blotch) than wheat fertilized with adequate nitrogen. Soil-applied copper increased yield and decreased incidence and severity of FHB in spring wheat in North Dakota (Franzen et al., 2008). In Georgia, the severity of *S. nodorum* blotch in winter wheat increased with increasing rates of phosphorus applied prior to planting (Cunfer et al., 1980). Results from experiments conducted in south Asia showed that spot blotch and tan spot of wheat were reduced when plots were fertilized with potassium (Regmi et al., 2002; Sharma et al., 2005) and nitrogen, potassium, phosphorus and calcium (Sharma and Duveiller, 2004; Sharma et al., 2006). In a pot experiment in Brazil, spot blotch severity was reduced by 44% in wheat grown in soil amended with silicon (Domiciano et al., 2010). Results from a similar experiment in the United Kingdom showed that severities of the following diseases of wheat were reduced by the incorporation of silicon into the potting medium: powdery mildew, *S. nodorum* blotch, *S. tritici* blotch and eyespot (Rodgers-Gray and Shaw, 2004).

4.5 Irrigation management

In regions or areas where moisture is limiting, irrigation is often applied to optimize wheat yields. However, irrigation also creates an environment that is favourable to disease development because moisture is the dominant factor in the development of many plant disease epidemics (Agrios, 2005). In moisture-stressed environments, disease risk in irrigated fields is increased where conservation tillage is practised. To minimize disease risk, irrigation should be managed such that the wheat crop canopy dries between irrigations.

The effects of moisture from irrigation on disease development in wheat are illustrated by the following studies on FHB. In a field experiment in North Carolina, Cowger et al. (2009) investigated the effects of increasing post-anthesis moisture duration on FHB incidence and severity and on FDK, DON and *Fusarium*-infected kernels in soft red winter wheat cultivars varying in FHB resistance and spray-inoculated with macroconidia of *F. graminearum* at mid-anthesis. Averaged across cultivars, 10 or 20 days of post-anthesis mist significantly increased FHB incidence and severity, FDK, DON and *Fusarium*-infected kernels compared to no mist. In Minnesota, Culler et al. (2007) compared the effect of standard versus extended mist irrigation on FDK and DON in *F. graminearum*-inoculated field plots of spring wheat. The percentage of FDK was higher under extended mist irrigation compared to standard irrigation. However, DON concentration was reduced under extended mist irrigation, and this was attributed to leaching of DON from spikes in the plots with extended irrigation. In field experiments in Austria, Lemmens et al. (2004) inoculated winter wheat cultivars at the anthesis growth stage with a spore suspension of *F. culmorum*, another *Fusarium* sp. that causes FHB. They simulated a wet growing season with a mist irrigation system that kept the crop canopy wet for nearly four weeks after flowering. Results showed that FHB severity was higher and spike weight was lower in the irrigated compared to the non-irrigated treatments.

© Burleigh Dodds Science Publishing Limited, 2017. All rights reserved.

4.6 Control of volunteer cereals and weeds

Volunteer cereals and weeds can serve as hosts for some wheat diseases. If volunteer cereals and weeds are not controlled, propagules of oversummering or overwintering pathogens can be dispersed from them to newly emerged wheat in the fall or spring. Examples of fungal pathogens that can oversummer on volunteer wheat and infect winter wheat in the fall are *Puccinia striiformis* f. sp. *tritici*, causal agent of stripe rust, and *Puccinia triticina*, causal agent of leaf rust. In 2014 and 2015, these two pathogens infected winter wheat in the Nebraska panhandle (northwest Nebraska) a few weeks after planting in the fall (personal observation). In mild winters, these pathogens can overwinter and cause early rust infections in the spring in both spring and winter wheat.

Examples of viruses that can survive on volunteer cereals and weeds and infect wheat in the fall or spring are barley yellow dwarf viruses (BYDV), vectored by aphids; and WSMV, *Triticum* mosaic virus and wheat mosaic virus, vectored by wheat curl mites. In addition to wheat, these viruses can infect wild and pasture grasses, as well as grassy weeds in crop fields, and the grasses also can serve as hosts for the vectors (Guy, 1988; Irwin and Thresh, 1990; Smith, 1963; Staples and Allington, 1956). Therefore, farmers should avoid planting wheat next to pasture grasses or in fields where volunteer wheat and grassy weeds are not controlled because of the high risk of infection by these viruses (Jones, 2013; Irwin and Thresh, 1990; Navia et al., 2013; Smith, 1963; Staples and Allington, 1956; Wegulo et al., 2008).

5 Chemical control

Since the development of systemic fungicides in the late 1960s, the use of fungicides in cereal production has become routine (Hewitt, 1998). The fungicides used to control wheat diseases belong to different chemical classes with unique modes of action against fungal pathogens. Two representative classes widely used as foliar fungicides in wheat are the triazoles and strobilurins (Wegulo et al., 2011b). Fungicides in both classes are also used as seed treatments. Triazoles have a five-membered ring of two carbon atoms and three nitrogen atoms. They move systemically through the plant xylem and are curative. They act by slowing fungal growth through the inhibition of sterol biosynthesis (Buchenauer, 1987). Sterols are essential building blocks of fungal cell membranes and are inhibited at a single site by triazoles. Because of their curative activity against early fungal infections and their ability to redistribute in the crop, triazoles are highly effective and reliable (Hewitt, 1998). Examples of triazoles used in cereal crop production are difenoconazole, metconazole, propiconazole, prothioconazole, tebuconazole and triticonazole.

The strobilurins are named in recognition of the mushroom *Strobilurus tenacellus*. They are quinone outside inhibitors and work by interfering with energy production in fungi (Hewitt, 1998). They act as local systemics by inhibiting fungal spore germination and early infection and are highly effective when applied preventively. The strobilurins have a single-site mode of action. Examples of strobilurin fungicides used in wheat are azoxystrobin, fluoxastrobin, picoxystrobin, pyraclostrobin and trifloxystrobin. Other chemical classes of fungicides used in wheat, mainly as seed treatments, include acylalanines, benzimidazoles, carboxamides, dithiocarbamates and phenylpyrroles (Hewitt, 1998).

© Burleigh Dodds Science Publishing Limited, 2017. All rights reserved.

5.1 Seed treatment fungicides

The first fungicides to be developed were seed treatments for control of cereal bunts and smuts (Hewitt, 1998; Mathre et al., 2001). Organomercury seed treatments were widely used from the 1920s to the 1970s and 1980s when they were banned due to their adverse toxicity (Hewitt, 1998; Mathre et al., 2001; Richardson, 1986). To date, several different chemical classes of seed treatment fungicides (see preceding paragraph) have been developed and are available commercially. Seed treatments control seed transmitted pathogens that may be surface-borne on the seed or internally seed-borne. Additionally, they control soil-borne pathogens. Systemic seed treatments also provide additional protection against foliar fungal diseases such as rusts that can occur after wheat emergence in the fall or early spring. Insecticide or fungicide–insecticide combination seed treatments control insect pests such as wireworms, Hessian fly and fall season aphids which transmit BYDV. By controlling seed and seedling diseases, seed treatments improve stand establishment and result in healthy, vigorous seedlings.

Wheat diseases that can be controlled with seed treatments include common bunt, dwarf bunt (*Tilletia controversa*), ergot (*Claviceps purpurea*), karnal bunt (*Tilletia indica*), loose smut (*Ustilago nuda*), seed decay and seedling blights (*Pythium* spp., *Penicillium* spp., *Fusarium* spp.), take-all and root and crown rots (*Rhizoctonia solani*, *Fusarium* spp., *B. sorokiniana*) (Mathre et al., 2001). The following examples illustrate the effectiveness of seed treatments in wheat. Efficacy ranges from low for some products to high for others. In Scotland, field trials were conducted over a five-year period to assess the need for routine use of organomercurial cereal seed treatment fungicides (Richardson, 1986). Establishment of winter wheat increased from 59% to 65% with seed treatment, and the level of seedling diseases in untreated wheat and barley crops was two to three times that in treated crops. However, seed treatments had no effect on yield. In field experiments in Washington State, thiabendazole applied as a seed treatment effectively controlled dwarf bunt and seed- and soil-borne common bunt. However, the rate at which it was effective varied with trial location (Hoffmann, 1971).

In field trials in Idaho, Montana, Utah and Washington State, difenoconazole applied as a seed treatment provided complete control of dwarf bunt in winter wheat (Sitton et al., 1993; Keener et al., 1995). Other seed treatments included in the study by Sitton et al. (1993) (thiabendazole, triadimenol and carboxin + thiram) did not provide adequate control. In the study by Keener et al. (1995), yield was similar in treated and untreated cultivars in the first year of the trial, but was 21% greater in treated than in non-treated cultivars in the second year. In Ontario, Canada, difenoconazole + metalaxyl-M (Dividend XL RTA) provided 98–99% control of dwarf bunt in winter wheat cultivars and germplasm lines (Xue et al., 2007). Two other products tested in the study, carbathiin + thiram (Vitaflo-280) and *Clonostachys rosea* (ACM941) provided 50% and 35% control, respectively. The seed treatments increased emergence by 5–8% and yield by 2–4%. In Australia, a new seed treatment fungicide, ipconazole + metalaxyl + N-methyl-2-pyrrolidone (Rancona Dimension), reduced *Rhizoctonia* root rot severity in wheat by up to 61% compared to 23% reduction by difenoconazole + metalaxyl-M (Dividend), the recommended standard seed treatment fungicide (Almasudy et al., 2015). Reductions in BYDV by insecticide seed treatments which reduce aphid vector populations have been demonstrated in Australia (McKirdy and Jones, 1996) and Ireland (Kennedy and Connery, 2012). Because seed treatments are relatively inexpensive but effective, farmers are encouraged to use fungicide-treated seed to avoid the risk of losing entire wheat crops, which often is the case when diseases like common bunt occur.

© Burleigh Dodds Science Publishing Limited, 2017. All rights reserved.

5.2 Foliar fungicides

Foliar fungicides are used to control above ground fungal diseases of wheat. These diseases include leaf rust, stripe rust, stem rust (*Puccinia graminis* f. sp. *tritici*), powdery mildew (*Blumeria graminis* f. sp. *graminis*), tan spot, *S. tritici* blotch, spot blotch (*C. sativus*), *S. nodorum* blotch and FHB. New fungicide chemistries have been and continue to be developed in part to increase efficacy and overcome the resistance to older chemistries in pathogen populations. The benefits of using fungicides are manifested in increased yields and therefore higher profits for the farmer. In the U.K., a yield response of up to 89% was realized from applying fungicides to winter wheat in experiments conducted from 1978 to 1982. The value of the net yield from fungicide application to cereals was double the fungicide costs (Cook and King, 1984). In Denmark, a net yield of up to 2700 kg ha^{-1} resulted from applying fungicides to control powdery mildew and *Septoria* diseases (Jørgensen et al., 2000). In Sweden, there was a mean net return of US$28 ha^{-1} from using fungicides in winter wheat during the period from 1995 to 2007 and during the period from 1983 to 2007 the mean net return was $16 ha^{-1} (Wiik and Rosenqvist, 2010).

In the U.S., disease reduction and yield increases in spring and winter wheat due to fungicide use have been demonstrated in various field trials. In multi-state trials, fungicide applications significantly reduced FHB and DON and increased yield (Willyerd et al., 2012; Wegulo et al., 2011a). In field experiments to determine the efficacy of fungicides in controlling foliar diseases of winter wheat, Wegulo et al. (2009) found that yield was up to 42% higher in fungicide-sprayed plots compared to non-sprayed plots. In a study to evaluate the effect of planting date, with or without fungicide application, on grain yield, yield components and grain quality, Kelley (2001) showed that over a period of six years, the fungicide propiconazole significantly increased winter wheat yield 77% of the time. Significant yield increases resulted from fungicide application to control the disease complex of leaf rust, tan spot and *S. tritici* blotch in winter wheat (Vamshidhar et al., 1998). In field experiments conducted by Ransom and McMullen (2008), fungicides improved yields by 5.5–44.0%. Tebuconazole applied at Zadoks growth stage (GS) 37 (Zadoks et al., 1974) and propiconazole applied at GS 37 followed by triadimefon + mancozeb at GS 55 effectively controlled leaf rust and *S. tritici* blotch, resulting in the lowest disease severities and highest winter wheat yields (Milus, 1994).

To be effective and economically profitable, foliar fungicide application should be timed to protect the wheat crop during the critical stages of growth. In the U.S., fungicides are usually applied to winter wheat one to three times per season. The most common application is timed at 50% to full flag leaf emergence. An earlier application during stem elongation to control early season diseases such as tan spot or a later application at anthesis to suppress FHB may be warranted depending on the level of disease risk.

6 Biological control

'Biological control is the reduction of inoculum density or disease-producing activities of a pathogen or parasite in its active or dormant state, by one or more organisms, accomplished naturally or through manipulation of the environment, host, or antagonist, or by mass introduction of one or more antagonists' (Baker and Cook, 1982). For biological control to be successful, the environment must be favourable to growth and multiplication of the antagonist (microorganism that interferes with growth and multiplication of the pathogen or

© Burleigh Dodds Science Publishing Limited, 2017. All rights reserved.

parasite). In wheat, a classic example of biological control is take-all decline, a phenomenon where take-all is severe during the first few years of a wheat monoculture, then declines in subsequent years (Baker and Cook, 1982). Take-all decline is attributed to the build-up, in the soil and wheat rhizosphere, of microorganisms that antagonize the take-all pathogen, *Ggt*. These microorganisms have been identified as bacteria, mainly fluorescent *Pseudomonas* spp. and actinomycetes (Baker and Cook, 1982; Cook and Baker, 1983; Weller and Cook, 1983). Many studies have demonstrated the ability of biological control agents (BCAs) to reduce root, stem base, foliar and spike diseases of wheat. The following examples illustrate the efficacy of BCAs in controlling or suppressing diseases of wheat.

Experiments conducted in China and Washington State under greenhouse conditions showed that fluorescent *Pseudomonas* spp. isolated from wheat in China and applied to wheat seed significantly reduced the severity of take-all and *Rhizoctonia* root rot (Yang et al., 2011, 2014). In an earlier study in Washington State, fluorescent *Pseudomonas* spp. isolated from the roots of wheat grown in soil that was naturally suppressive to take-all were applied to seed of winter and spring wheat. The bacteria significantly reduced take-all severity in greenhouse and field tests (Weller and Cook, 1983). In field trials in Idaho, the biofumigant fungus *M. albus*, when applied as a seed or an in furrow soil treatment, reduced common bunt incidence in spring wheat by up to 35% (Goates and Mercier, 2011).

Recent research efforts on biological control have focused on FHB, a disease that can be controlled only partially with fungicides and cultural practices. Among the BCAs that have been identified and tested in the greenhouse and field for FHB control are bacteria and fungi. Among bacteria, antagonistic activity against FHB-causing *Fusarium* spp. has been demonstrated for *Bacillus* spp. (Schisler et al., 2002; Zhao et al., 2014), *Pseudomonas* spp. (Schisler et al., 2006), *Lysobacter enzymogenes* (Jochum et al., 2006) and *Streptomyces* spp. (Palazzini et al., 2007). Fungal antagonists include *Cryptococcus* spp. (Schisler et al., 2011), *Trichoderma* spp. (Matarese et al., 2012), *C. rosea* (Xue et al., 2014a,b) and *Aureobasidium pullulans* (Wachowska and Glowacka, 2014). These BCAs can be applied directly to spikes to slow down disease progression (Xue et al., 2014b) or to residues to suppress the production of perithecia (Xue et al., 2014a).

Although the antagonistic activity of BCAs against wheat diseases has been demonstrated in experiments, little progress has been made in developing them to quantities and formulations that can be used on a commercial scale. This is in part due to the time needed to screen hundreds or thousands of microorganisms to identify effective antagonists, difficulty in determining the right media and environmental conditions for culturing identified candidates to raise sufficient populations for greenhouse and field trials, and challenges in formulating carrier media that can preserve biomass and prolong shelf life (Schisler et al., 2004). Significant progress has been made in Canada where *C. rosea* has been formulated to a product that is effective in reducing perithecial production on crop residues by *Gibberella zeae* (sexual stage of *F. graminearum*) and in suppressing FHB in the field (Xue et al., 2014a, b). A biopesticide company has developed the product into a biofungicide that will soon be commercially available to wheat growers (Forsythe, 2015).

7 Use of disease forecasting systems

Forecasting systems use input variables such as weather and disease resistance levels of wheat varieties to calculate the risk of disease outbreaks. Hence, growers can deploy

© Burleigh Dodds Science Publishing Limited, 2017. All rights reserved.

management tactics only when necessary. Because fungicide application in the absence of damaging levels of disease is not profitable and can result in a net loss for the grower (Wegulo et al., 2011c), use of disease forecasting systems saves growers time and money. FHB is used as an example to illustrate the utility of disease forecasting systems in managing wheat diseases. Due to more frequent and severe outbreaks of FHB in North America and other countries during the last two decades, several forecasting systems have been developed to optimize the use of fungicide applications to suppress the disease and the associated mycotoxin DON. In the U.S., the FHB Risk Assessment Tool (http://www.wheatscab.psu.edu/) estimates the risk of FHB outbreaks using weather variables during the previous seven days. FHB risk is predicted with over 75% accuracy. This forecasting system is deployed in more than 20 states where wheat and barley are grown in the U.S. In Canada, the DONcast® model (http://www.weatherinnovations. com/doncast.cfm) enables growers to make efficient fungicide application decisions by predicting DON accumulation at full spike emergence. Inputs into the model include weather variables, cultivar, crop rotation and tillage type. In Switzerland, a similar forecasting system, FusaProg (Musa et al., 2007), was developed to predict the risk of field-specific DON accumulation based on weather variables, crop rotation, tillage type and cultivar. Growers who use these models save time and money by avoiding unnecessary application of fungicides. By applying fungicides when necessary based on warnings issued by the disease forecasting systems, losses due to FHB and DON are reduced and better quality grain is made available to the food and malting industries. In North Dakota, the Small Grain Disease Forecasting Model (https://www.ag.ndsu.edu/ cropdisease) estimates the risk of foliar diseases as well.

8 Integrated disease management

The most effective approach to managing wheat diseases is to integrate as many available tactics as practically possible into an integrated disease management (IDM) programme. Previous research has shown that integrating two or more management tactics is more effective in reducing wheat diseases than using one tactic. In experiments conducted in Kansas and Nebraska, Wegulo et al. (2011a) found that integrating cultivar resistance and fungicide application was more effective in reducing FHB and DON in winter wheat than using either tactic alone. In a year with high disease pressure, fungicide efficacy in reducing FHB index was 66% in the most resistant cultivar compared to 19% in the most susceptible cultivar. Efficacy in reducing DON was 62% in the most resistant cultivar compared to 20% in the most susceptible cultivar. Greater effectiveness in reducing FHB and DON by combining resistance with fungicide application was similarly demonstrated by Willyerd et al. (2012), McMullen et al. (2008) and Bockus et al. (2011–15). In North Dakota, McMullen et al. (2008) found that rotation alone, rotation + a tolerant cultivar and rotation + a tolerant cultivar + fungicide application reduced FHB by 50%, 80% and 92%, respectively.

Research on other pathosystems also has demonstrated the superior effectiveness of integrating two or more tactics to manage wheat diseases. Goates and Mercier (2011) showed that control of common bunt by the biofumigant fungus *M. albus* applied as a seed or an in furrow treatment in spring wheat in Idaho was more effective at a later than an earlier planting date. They attributed the greater effectiveness at the later

© Burleigh Dodds Science Publishing Limited, 2017. All rights reserved.

planting date to warmer soil temperatures that were not favourable to seedling infection by the common bunt fungus. Epidemiological studies conducted in Nebraska indicated that the most effective control strategy for WSMV was integration of volunteer wheat control, late planting and elimination of grasses other than wheat (Staples and Allington, 1956). Late planting, however, can have adverse effects on yield. Therefore, the current recommendation in Nebraska is to avoid early planting but plant at the recommended date for a given location (Wegulo et al., 2008).

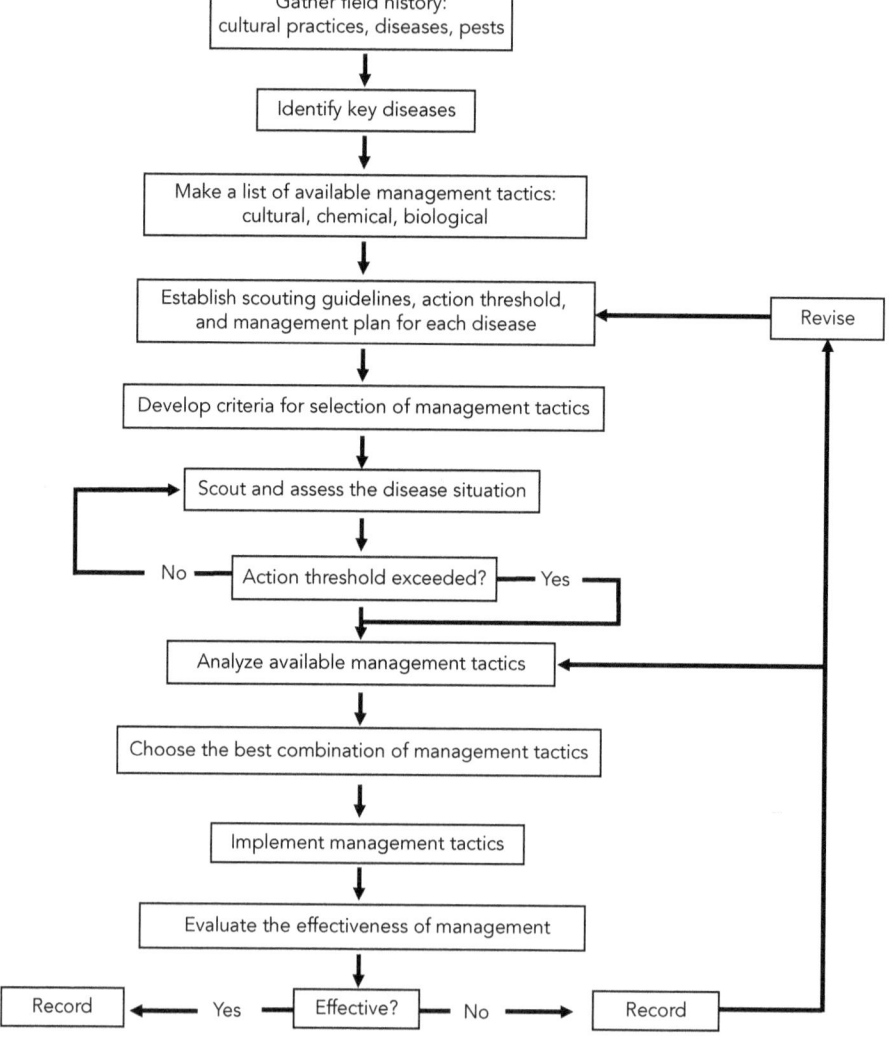

Figure 1 A flow chart depicting the implementation of an effective integrated disease management (IDM) programme. Adapted and modified from Hamm et al. (1990).

© Burleigh Dodds Science Publishing Limited, 2017. All rights reserved.

A successful IDM programme for wheat depends on well-thought-out and planned procedures for each disease and situation. The detailed components of the programme will vary with geographical area due to differences in the predominant diseases among areas or regions. The following steps for developing an IDM programme for wheat, summarized in the flow chart in Fig. 1, are adapted and modified from Flint and Gouveia (2001). (i) Based on farm history, identify the key diseases that occur in your wheat fields. (ii) Establish scouting guidelines for each disease. Scouting should begin as early in the growing season as practical for detection of early season diseases such as residue-borne diseases and stripe rust. (iii) Establish the action threshold for each disease. Wheat diseases are best controlled preventively. For diseases that can be controlled by spraying a fungicide, the decision to spray is dependent on disease detection, disease-favourable environmental conditions and crop growth stage. (iv) Make a list of management strategies for each disease. Some management strategies are long term, whereas others are short term. (v) For each disease, develop specific criteria for selection of management tactics. For some diseases such as wheat streak mosaic, management tactics are implemented before planting, whereas for other diseases, they are implemented during the growing season. (vi) For diseases that can be controlled with seed treatment or foliar fungicides, have guidelines for selecting the most effective fungicides. Information on fungicide efficacy can be obtained from extension personnel and publications. If fungicides are applied more than once during the growing season, rotate the chemical class (mode of action) to prevent build-up of fungicide resistance in pathogen populations. (vii) Have in place a thorough record-keeping system. Records serve as an invaluable reference resource and help in evaluation and improvement of the IDM programme. (viii) Make a list of resources from which you can obtain help when you need it. Useful resources include clinics for disease identification, university personnel and publications and private consultants. (ix) Build flexibility into the IDM programme to deal with changing disease situations. Diseases and their intensity can vary from season to season depending on the weather. (x) For diseases for which disease forecasting systems have been developed, use information from these systems to make management decisions. (xi) Monitor the weather and disease information from the university or government extension service and commercial news outlets. Know the disease situation in surrounding areas or regions. (xii) In addition to disease management, implement agronomic practices (variety selection, crop rotation, tillage type, fertilization, weed and insect control, irrigation) that will maximize productivity of your wheat crop.

Educational programmes that enhance the knowledge of farmers and others involved in IDM are essential for success. These programmes should be offered on a regular basis and include farmers field schools which have been used widely to train farmers in resource-limited countries (El Khoury and Makkouk, 2010), demonstrations, field days, workshops and clinics. Content delivered during these programmes should emphasize the general principles of IDM, specific disease management tactics, practical aspects of disease management including personal and environmental safety and updates on new products and information that can be used to more effectively manage wheat diseases.

9 Future trends in research

There is a need for concerted efforts to develop new integrated disease management strategies or improve on those currently in use for more effective and sustainable

© Burleigh Dodds Science Publishing Limited, 2017. All rights reserved.

management of wheat diseases. New sources of disease resistance are needed for introgression into wheat cultivars with desirable agronomic traits. Investment in research and development (R&D) for new fungicide chemistries that can be used in rotation to control wheat diseases will result in more effective disease control and will mitigate the risk of fungicide resistance development in pathogen populations. Surveys to detect new pathogens as well as shifts in known pathogen species will provide information that will be invaluable in managing wheat diseases. Improvement in the accuracy and robustness of disease forecasting systems and increased adoption of these systems by farmers will lead to more effective disease management. Commercialization of biological control agents will increase the options available for integrated disease management. Improvement in fungicide spray technology will increase coverage, retention and efficacy of fungicides applied to wheat foliage and spikes. The ultimate goal of these efforts is increased and sustainable production of high-quality wheat that is needed to feed an increasing and hungry world population.

10 Where to look for further information

Further information on integrated wheat disease management can be obtained from various sources. The extension service in institutions of higher learning and governments can provide information that is locally relevant. Many institutions of higher learning that have an extension service avail this information through newsletters and extension publications such as guides and circulars that may be accessed freely on the worldwide web or purchased for a small fee. They also have specialists and educators who can be called upon to provide information locally. Other sources of information include textbooks, refereed journal papers, agricultural news magazines, radio and television stations devoted to agricultural news and private consultants.

11 References

Agrios, G. N. (2005), *Plant Pathology Fifth Edition*, Academic Press, Amsterdam.

Almasudy, A. M., You, M. P. and Barbetti, M. J. (2015), 'Influence of fungicidal seed treatments and soil type on severity of root disease caused by Rhizoctonia solani AG-8 on wheat', *Crop Prot.*, 75: 40–5.

Angus, J. F., Kirkegaard, J. A., Hunt, J. R., Ryan, M. H., Ohlander, L. and Peoples, M. B. (2015), 'Break crops and rotations for wheat', *Crop Pasture Sci.*, 66: 523–52.

Anonymous (1968), *Plant Disease Development and Control, Vol. I, Principles of Plant and Animal Pest Control*, publ. 1596, Nat. Acad. Sci., Washington, DC.

Anonymous (1990), 'Tillage definitions', *Conserv. Impact*, 8(10): 7.

Anonymous (1994), 'Integrated pest management practices on fruit and nuts', *RTD Updates: Pest Management*, USDA-ERS.

Bailey, K. L. and Duczek, L. J. (1996), 'Managing cereal diseases under reduced tillage', *Can. J. Plant Pathol.*, 18(2): 159–67.

Bailey, K. L., Mortensen, K. and Lafond, G. P. (1992), 'Effects of tillage systems and crop rotations on root and foliar diseases of wheat, flax, and peas in Saskatchewan', *Can. J. Plant Sci.*, 72: 583–91.

Baker, K. F. and Cook, R. J. (1982), *Biological Control of Plant Pathogens*, The American Phytopathological Society, St. Paul.

© Burleigh Dodds Science Publishing Limited, 2017. All rights reserved.

Barkley, A. P. and Porter, L. L. (1996), 'The determinants of wheat variety selection in Kansas, 1974 to 1993', *Amer. J. Agr. Econ.*, 78(2): 202–11.

Bockus, W. W. (1998), 'Control strategies for stubble-borne pathogens of wheat', *Can. J. Plant Pathol.*, 20(4): 371–5.

Bockus, W. W. and Claassen, M. M. (1992), 'Effects of crop rotation and residue management practices on severity of tan spot of winter wheat', *Plant Dis.*, 76(6): 633–6.

Bockus, W. W., Davis, M. A. and Norman, B. L. (1994), 'Effect of soil shading by surface residues during summer fallow on take-all of winter wheat', *Plant Dis.*, 78(1): 50–4.

Bockus, W. W., De Wolf, E. D. and Wegulo, S. N. (2011), 'Effect of foliar fungicide application on Fusarium head blight in eight winter wheat cultivars, 2010', *Plant Dis. Manag. Rep.*, 5: CF009.

Bockus, W. W., De Wolf, E. D. and Wegulo, S. N. (2012), 'Effect of foliar fungicide application on Fusarium head blight in eight winter wheat cultivars, 2011', *Plant Dis. Manag. Rep.*, 6: CF004.

Bockus, W. W., De Wolf, E. D. and Wegulo, S. N. (2013), 'Effect of foliar fungicide application on Fusarium head blight in eight winter wheat cultivars, 2012', *Plant Dis. Manag. Rep.*, 7: CF018.

Bockus, W. W., De Wolf, E. D. and Wegulo, S. N. (2014), 'Effect of foliar fungicide application on Fusarium head blight in eight winter wheat cultivars, 2013', *Plant Dis. Manag. Rep.*, 8: CF005.

Bockus, W. W., De Wolf, E. D. and Wegulo, S. N. (2015), 'Effect of Prosaro fungicide application on Fusarium head blight in eight winter wheat cultivars, 2014', *Plant Dis. Manag. Rep.*, 9: CF003.

Bockus, W. W., O'Connor, J. P. and Raymond, P. J. (1983), 'Effect of residue management method on incidence of Celphalosporium stripe under continuous winter wheat production', *Plant Dis.*, 67(12): 1323–4.

Bockus, W. W. and Shroyer, J. P. (1998), 'The impact of reduced tillage on soilborne plant pathogens', *Annu. Rev. Phytopathol.*, 36: 485–500.

Buchenauer, H. (1987), 'Mechanism of action of triazolyl fungicides and related compounds', pages 205–31, in *Modern Selective Fungicides: Properties, Applications, Mechanisms of Action*, H. Lyr (Ed.), Wiley, New York.

Carris, L. M. (2010), 'Common bunt (stinking smut)', pages 60–2, in *Compendium of Wheat Diseases and Pests*, third edition, W. W. Bockus, R. L. Bowden, R. M. Hunger, W. L. Morrill, T. D. Murray and R. W. Smiley (Eds), APS Press, St. Paul.

Chen, Y., Zhang, F., Tang, L., Zheng, Y., Li, Y., Christie, P. and Li, L. (2007), 'Wheat powdery mildew and foliar N concentrations as influenced by N fertilization and belowground interactions with intercropped faba bean', *Plant and Soil*, 291(1–2): 1–13.

Cook, R. J. (2003), 'Take-all of wheat', *Physiol. Mol. Plant Pathol.*, 62(2): 73–86.

Cook, R. J. and Baker, K. F. (1983), *The Nature and Practice of Biological Control of Plant Pathogens*, The American Phytopathological Society, St. Paul.

Cook, R. J. and King, J. E. (1984), 'Loss caused by cereal diseases and the economics of fungicidal control', pages 237–45, in *Plant Diseases: Infection, Damage and Loss*, Blackwell, Oxford.

Cowger, C., Patton-Özkurt, J., Brown-Guedira, G. and Perugini, L. (2009), 'Post-anthesis moisture increased Fusarium head blight and deoxynivalenol levels in North Carolina winter wheat', *Phytopathology*, R. K. S. Wood and G. J. Jellis (Eds), 99(4): 320–7.

Cromey, M. G., Francis, G. S., Trimmer, L. A., Tably, F. J., Gillespie, R. N., Fraser, P. M., Pearson, A. J., Butler, R. C., Curtin, D. and Bithell, S. L. (2008), 'Influence of crop rotation, tillage, residue management and winter cover crop on take-all in spring wheat', *N.Z. Plant Protect.*, 61: 261–9.

Culler, M. D., Miller-Garvin, J. E. and Dill-Macky, R. (2007), 'Effect of extended irrigation and host resistance on deoxynivalenol accumulation in *Fusarium*-infected wheat', *Plant Dis.*, 91(11): 1464–72.

Cunfer, B. M., Buntin, G. D. and Phillips, D. V. (2006), 'Effect of crop rotation on take-all of wheat in double cropping systems', *Plant Dis.*, 90(9): 1161–6.

Cunfer, B. M., Touchton, J. T. and Johnson, J. W. (1980), 'Effects of phosphorus and potassium fertilization on Septoria glume blotch of wheat', *Phytopathology*, 70(12): 1196–9.

© Burleigh Dodds Science Publishing Limited, 2017. All rights reserved.

Daroub, S. H. and Snyder, G. H. (2007), 'The chemistry of plant nutrients in soil', pages 1–8, in *Mineral Nutrition and Plant Disease*, L. E. Datnoff, W. H. Elmer and D. M. Huber (Eds), APS Press, St. Paul.

Datnoff, L. E., Elmer, W. H. and Huber, D. M. (Eds) (2007), *Mineral Nutrition and Plant Disease*, APS Press, St. Paul.

Detlefsen, N. K. and Jensen, A. L. (2004), 'A stochastic model for crop variety selection', *Agr. Syst.*, 81(1): 55–72.

Devadas, R., Lamb, D. W., Backhouse, D. and Simpfendorfer, S. (2015), 'Sequential application of hyperspectral indices for delineation of stripe rust infection and nitrogen deficiency in wheat', *Precision Agric.*, 16(5): 477–91.

Dill-Macky, R. (2010), 'Fusarium head blight (scab)', pages 34–6, in *Compendium of Wheat Diseases and Pests*, third edition, W. W. Bockus, R. L. Bowden, R. M. Hunger, W. L. Morrill, T. D. Murray and R. W. Smiley (Eds), APS Press, St. Paul.

Dill-Macky, R. and Jones, R. K. (2000), 'The effect of previous crop residues and tillage on Fusarium head blight of wheat', *Plant Dis.*, 84(1): 71–6.

Domiciano, G. P., Rodrigues, F. A., Vale, F. X. R., Filha, M. S. X., Moreira, W. R., Andrade, C. C. L. and Pereira, S. C. (2010), 'Wheat resistance to spot blotch potentiated by silicon', *J. Phytopathol.*, 158(5): 334–43.

El Khoury, W. and Makkouk, K. (2010), 'Integrated plant disease management in developing countries', *J. Plant Pathol.*, 92(4): S4.35–S4.42.

Ennaïfar, S., Lucas, P., Meynard, J.-M. and Makowski, D. (2005), 'Effects of summer fallow management on take-all of winter wheat caused by *Gaeummanomyces graminis* var. *tritici*', *Eur. J. Plant Pathol.*, 112: 167–81.

Fenster, C. R., Boosalis, M. G. and Weihing, J. L. (1972), *Date of Planting Studies of Winter Wheat and Winter Barley in Relation to Root and Crown Rot and Grain Yields and Quality*, Research Bulletin No. 250, University of Nebraska-Lincoln College of Agriculture, The Agricultural Experiment Station.

Fischer, R. A. (2009), 'Foreword', pages xvii–xviii, in *Wheat: Science and Trade*, B. F. Carver (Ed.), Wiley-Blackwell, Hoboken.

Flint, M. L. and Gouveia, P. (2001), *IPM in Practice: Principles and Methods of Integrated Pest Management*, University of California Publication 3418. Oaklnad, CA.

Forsythe, T. K. (2015), 'Biofungicide for Fusarium head blight of wheat made in Canada', Ag Annex, online: http://www.agannex.com/diseases/biofungicide-for-fusarium-head-blight-of-wheat-made-in-canada, accessed on 30 October 2015.

Franzen, D. W., McMullen, M. V. and Mosset, D. S. (2008), 'Spring wheat and durum yield and disease responses to copper fertilization of mineral soils', *Agron. J.*, 100(2): 371–5.

Goates, B. J. and Mercier, J. (2011), 'Control of common bunt of wheat under field conditions with the biofumigant fungus Muscodor albus', *Eur. J. Plant Pathol.*, 131(3): 403–7.

Graham, R. D. (1983), 'Effects of nutrient stress on susceptibility of plants to disease with particular reference to the trace elements', pages 221–76, in *Advances in Botanical Research Vol. 10*, H. W. Woodhouse (Ed.), Academic Press, New York.

Gray, S. M. (2010), 'Barley yellow dwarf', pages 100–2, in *Compendium of Wheat Diseases and Pests*, third edition, W. W. Bockus, R. L. Bowden, R. M. Hunger, W. L. Morrill, T. D. Murray and R. W. Smiley (Eds), APS Press, St. Paul.

Gurung, S., Sharma, R. C., Duveiller, E. and Shrestha, S. M. (2012), 'Comparative analyses of spot blotch and tan spot epidemics on wheat under optimum and late sowing period in south Asia', *Eur. J. Plant Pathol.*, 134(2): 257–66.

Guy, P. L. (1988), 'Pasture ecology of barley yellow dwarf viruses at Sandford, Tasmania', *Plant Pathol.*, 37(4): 546–50.

Hamm, P. B., Campbell, S. J. and Hansen, E. M. (1990), *Growing Healthy Seedlings: Identification and Management of Pests in Northwest Forest Nurseries*, Forest Pest Management, U.S. Dept. Agric. And Forest Research Laboratory, Oregon State University.

© Burleigh Dodds Science Publishing Limited, 2017. All rights reserved.

Hesler, L. S., Riedell, W. E., Langham, M. A. C. and Osborne, S. L. (2005), 'Insect infestations, incidence of viral plant diseases, and yield of winter wheat in relation to planting date in the northern Great Plains', *J. Econ. Entomol.*, 98(6): 2020–7.

Hewitt, H. G. (1998), *Fungicides in Crop Protection*, CAB International, Wallingford.

Hoffmann, J. A. (1971), 'Control of common bunt of wheat by seed treatment with thiabendazole', *Phytopathology*, 61(9): 1071–4.

Hollier, C. A. and Hershman, D. E. (2008), 'Integrated pest management', pages 437–45, in *Plant Pathology Concepts and Laboratory Exercises*, R. K. Trigiano, M. T. Windham and A. S. Windham (Eds), CRC Press, New York.

Huber, D. M. (1991), 'The use of fertilizers and organic amendments in the control of plant disease', pages 357–94, in *Handbook of Pest Management in Agriculture Volume I*, CRC Press, Boca Raton.

Huber, D. M., Painter, C. G., McKay, H. C. and Peterson, D. L. (1968), 'Effect of nitrogen fertilization on take-all of winter wheat', *Phytopathology*, 58(12): 1470–2.

Huber, D. M. and Watson, R. D. (1974), 'Nitrogen form and plant disease', *Annu. Rev. Phytopathol.*, 12: 139–65.

Hunger, R. M. (2010), 'Wheat streak mosaic', pages 115–7, in *Compendium of Wheat Diseases and Pests*, third edition, W. W. Bockus, R. L. Bowden, R. M. Hunger, W. L. Morrill, T. D. Murray and R. W. Smiley (Eds), APS Press, St. Paul.

Irwin, M. E. and Thresh, J. M. (1990), 'Epidemiology of barley yellow dwarf: a study in ecological complexity', *Annu. Rev. Phytopathol.*, 28: 393–424.

Jochum, C. C., Osborne, L. E. and Yuen, G. Y. (2006), 'Fusarium head blight biological control with Lysobacter enzymogenes strain C3', *Biol. Control*, 39(3): 336–44.

Jones, R. A. (2013), 'Virus diseases of pasture grasses in Australia: incidences, losses, epidemiology, and management', *Crop Pasture Sci.*, 64(3): 216–33.

Jørgensen, L. N., Henriksen, K. E. and Nielsen, G. C. (2000), 'Margin over cost in disease management in winter wheat and spring barley in Denmark', pages 655–62, in *Brighton Crop Protection Conference: Pests & Diseases – 1998: Volume 2: Proceedings of an International Conference, Brighton*, UK, 13–16 November 2000.

Keener, T. K., Stougaard, R. N. and Mathre, D. E. (1995), 'Effect of winter wheat cultivar and difenoconazole seed treatment on dwarf bunt', *Plant Dis.*, 79(6): 601–4.

Kelley, K. W. (2001), 'Planting date and foliar fungicide effects on yield components and grain traits of winter wheat', *Agron. J.*, 93(2): 380–9.

Kennedy, T. F. and Connery, J. (2012), 'Control of barley yellow dwarf virus in minimum-till and conventional-till autumn-sown cereals by insecticide seed and foliar spray treatments', *J. Agric. Sci.*, 150(2): 249–62.

Khonga, E. B. and Sutton, J. C. (1988), 'Inoculum production and survival of Gibberella zeae in maize and wheat residues', *Can. J. Plant Pathol.*, 10(3): 232–9.

Kirkegaard, J., Christen, O., Krupinsky, J. and Layzell, D. (2008), 'Break crop benefits in temperate wheat production', *Field Crop. Res.*, 107(3): 185–95.

Krupinsky, J. M., Bailey, K. L., McMullen, M. P., Gossen, B. D. and Turkington, T. K. (2002), 'Managing plant disease risk in diversified cropping systems', *Agron. J.*, 94(2): 198–209.

Latin, R. X., Harder, R. W. and Wiese, M. V. (1982), 'Incidence of Cephalosporium stripe as influenced by winter wheat management practices', *Plant Dis.*, 66(3): 229–30.

Lemmens, M., Buerstmayr, H., Krska, R., Schuhmacher, R., Grausgruber, H. and Ruckenbauer, P. (2004), 'The effect of inoculation treatment and long-term application of moisture on Fusarium head blight symptoms and deoxynivalenol contamination in wheat grains', *Eur. J. Plant Pathol.*, 110: 299–308.

Maiorano, A., Blandino, M., Reyneri, A. and Vanara, F. (2008), 'Effects of maize residues on the *Fusarium* spp. infection and deoxynivalenol (DON) contamination of wheat grain', *Crop Prot.*, 27(2): 182–8.

© Burleigh Dodds Science Publishing Limited, 2017. All rights reserved.

Matarese, F., Sarrocco, S., Gruber, S., Seidl-Seiboth, V. and Vannacci, G. (2012), 'Biocontrol of Fusarium head blight: interactions between Trichoderma and mycotoxigenic *Fusarium*', *Microbiology*, 158(1): 98–106.

Mathre, D. E., Johnston, R. H. and Grey, W. E. (2001), 'Small grain cereal seed treatment', *The Plant Health Instructor*. DOI: 10.1094/PHI-I-2001-1008-01.

McKirdy, S. J. and Jones, R. A. C. (1996), 'Use of imidacloprid and newer generation synthetic pyrethroids to control the spread of barley yellow dwarf luteovirus in cereals', *Plant Dis.*, 80(8): 895–901.

McMullen, M., Halley, S., Schatz, B., Meyer, S., Jordahl, J. and Ransom, J. (2008), 'Integrated strategies for Fusarium head blight management in the United States', *Cereal Res. Commun.*, 36(Suppl.B.45): 563–8.

Milus, E. A. (1994), 'Effect of foliar fungicides on disease control, yield and test weight of soft red winter wheat', *Crop Prot.*, 13(4): 291–5.

Mundt, C. C. (2002), 'Use of multiline cultivars and cultivar mixtures for disease management', *Annu. Rev. Phytopathol.*, 40: 381–410.

Mundt, C. C. (2010), 'Cephalosporium stripe', pages 23–6, in *Compendium of Wheat Diseases and Pests*, third edition, W. W. Bockus, R. L. Bowden, R. M. Hunger, W. L. Morrill, T. D. Murray and R. W. Smiley (Eds), APS Press, St. Paul.

Murray, T. D., Parry, D. W. and Cattlin, N. D. (1998), *A Color Handbook of Diseases of Small Grain Cereal Crops*, CRC Press, New York.

Musa, T., Hecker, A., Vogelgsang, S. Forrer, H. R. (2007), 'Forecasting of Fusarium head blight and deoxynivalenol content in winter wheat with FusaProg.', *OEPP/EPPO Bulletin*, 37(2): 283–9.

Navia, D., de Mendonca, R. S., Skoracka, A., Szydlo, W., Knihinicki, D., Hein, G. L., da Silva Pereira, P. R. V., Truol, G. and Lau, D. (2013), 'Wheat curl mite, Aceria tosichella, and transmitted viruses: an expanding pest complex affecting cereal crops', *Exp. Appl. Acarol.*, 59(1–2): 95–143.

Nelson, P. and Meikle, S. (2001), 'NIAB interactive cereal variety Gross Margin Model: a decision support tool for UK farmers', *Farm Manag.*, 11(2): 130–3

Oerke, E. C. (2006), 'Crop losses to pests', *J. Agri. Sci.*, 144(1): 31–43.

Ogle, H. and Dale, M. (1997), 'Disease management: cultural practices', pages 390–404, in *Plant Pathogens and Plant Diseases*, Rockvale Publications, Armidale.

Palazzini, J. M., Ramirez, M. L., Torres, A. M. and Chulze, S. N. (2007), 'Potential biocontrol agents for Fusarium head blight and deoxynivalenol production in wheat', *Crop Prot.*, 26(11): 1702–10.

Paulitz, T. C. (2010), 'Take-all', pages 79–82, in *Compendium of Wheat Diseases and Pests*, third edition, W. W. Bockus, R. L. Bowden, R. M. Hunger, W. L. Morrill, T. D. Murray and R. W Smiley (Eds), APS Press, St. Paul.

Paulitz, T. C., Schroeder, K. L. and Schillinger, W. F. (2010), 'Soilborne pathogens of cereals in an irrigated cropping system: effects of tillage, residue management, and crop rotation', *Plant Dis.*, 94(1): 61–8.

Piccinni, G., Shriver, J. M. and Rush, C. M. (2001), 'Relationship among seed size, planting date, and common root rot in hard red winter wheat', *Plant Dis.*, 85(9): 973–6.

Plumb, R. T. and Johnstone, G. R. (1995), 'Cultural, chemical, and biological methods for the control of barley yellow dwarf', pages 307–19, in *Barley Yellow Dwarf: 40 years of Progress*, C. J. D'Arcy and P. A. Burnett (Eds), APS Press, St. Paul.

Ransom, J. K. and McMullen, M. P. (2008), 'Yield and disease control on hard winter wheat cultivars with foliar fungicides', *Agron. J.*, 100(4): 1130–7.

Rees, R. G., Platz, G. J. and Mayer, R. J. (1982), 'Yield losses in wheat from yellow spot: comparison of estimates derived from single tillers and plots', *Aust. J. Agr. Res.*, 33: 899–908.

Regmi, A. P., Ladha, J. K., Pasuquin, E., Pathak, H., Hobbs, P. R., Shrestha, L. L., Gharti, D. B. and Duveiller, E. (2002), 'The role of potassium in sustaining yields in a long-term rice-wheat experiment in the Indo-Gangetic plains of Nepal', *Biol. Fertil. Soils*, 36(3): 240–7.

© Burleigh Dodds Science Publishing Limited, 2017. All rights reserved.

Reis, E. M., Cook, R. J. and McNeal, B. L. (1982), 'Effect of mineral nutrition on take-all of wheat', *Phytopathology*, 72(2): 224–9.

Rice, R. W. (2007), 'The physiological role of minerals in the plant', pages 9–30, in *Mineral Nutrition and Plant Disease*, L. E. Datnoff, W. H. Elmer and D. M. Huber (Eds), APS Press, St. Paul.

Richardson, M. J. (1986), 'An assessment of the need for routine use of organomercurial cereal seed treatment fungicides', *Field Crop. Res.*, 13: 3–24.

Rodgers-Gray, B. S. and Shaw, M. W. (2004), 'Effects of straw and silicon soil amendments on some foliar and stem-base diseases in pot-grown winter wheat', *Plant Pathol.*, 53(6): 733–40.

Schisler, D. A., Khan, N. I., Boehm, M. J., Lipps, P. E., Slininger, P. J. and Zhang, S. (2006), 'Selection and evaluation of the potential of choline-metabolizing microbial strains to reduce Fusarium head blight', *Biol. Control,* 39(3): 497–506.

Schisler, D. A., Khan, N. I., Boehm, M. J. and Slininger, P. J. (2002), 'Greenhouse and field evaluation of biological control of Fusarium head blight on durum wheat', *Plant Dis.*, 86(12): 1350–6.

Schisler, D. A., Slininger, P. J., Behle, R. W. and Jackson, M. A. (2004), 'Formulation of *Bacillus* spp. for biological control of plant diseases', *Phytopathology*, 94(11): 1267–71.

Schisler, D. A., Slininger, P. J., Boehm, M. J. and Paul, P. A. (2011), 'Co-culture of yeast antagonists of Fusarium head blight and their effect on disease development in wheat', *Plant Pathology J.*, 10(4): 128–37.

Shaner, G. (1981), 'Effect of environment on fungal leaf blights of small grains', *Annu. Rev. Phytopathol.*, 19: 273–96.

Sharma, R. C. and Duveiller, E. (2004), 'Effect of helminthosporium leaf blight on performance of timely and late-seeded wheat under optimal and stressed levels of soil fertility and moisture', *Field Crops Res.*, 89(2–3): 205–18.

Sharma, S., Duveiller, E., Basnet, R., Karki, C. and Sharma, R. C. (2005), 'Effect of potash fertilization on Helmithosporium leaf blight severity in wheat, and associated increases in grain yield and kernel weight', *Field Crops Res.*, 93(2–3): 142–50.

Sharma, P., Duveiller, E. and Sharma, R. C. (2006), 'Effect of mineral nutrients on spot blotch severity in wheat, and associated increases in grain yield', *Field Crops Res.*, 95(2–3): 426–30.

Sieling, K., Ubben, K. and Christen, O. (2007), 'Effects of preceding crop, sowing date, N fertilization and fluquinconazole seed treatment on wheat growth, grain yield and take-all', *J. Plant Dis. Protect.*, 114(5): 213–20.

Simmons, F. W. and Nafziger, E. D. (2009), 'Soil Management and Tillage', pages 133–42, in *Illinois Agronomy Handbook, 24th Edition*, University of Illinois Extension, available online: http://extension.cropsciences.illinois.edu/handbook/pdfs/chapter10.pdf.

Sitton, J. W., Line, R. F., Waldher, J. T. and Goates, B. J. (1993), 'Difenoconazole seed treatment for control of dwarf bunt of winter wheat', *Plant Dis.*, 77(11): 1148–51.

Smiley, R. W. (2009), 'Water and temperature parameters associated with winter wheat diseases caused by soilborne pathogens', *Plant Dis.*, 93(1): 73–80.

Smiley, R. W., Backhouse, D., Lucas, P. and Paulitz, T. C. (2009), 'Diseases which challenge global wheat production-root, crown, and culm rots', pages 125–53, in *Wheat: Science and Trade*, B. F. Carver (Ed.), Wiley-Blackwell, Hoboken.

Smith, H. C. (1963), 'Control of barley yellow dwarf virus in cereals', *N.Z. J. Agric. Res.*, 6(3–4): 229–44.

Staples, R. and Allington, W. B. (1956), *Streak Mosaic of Wheat in Nebraska and its Control*, Research Bulletin No. 178, University of Nebraska-Lincoln College of Agriculture, Agricultural Experiment Station.

Subedi, K. D., Ma, B. L. and Xue, A. G. (2007), 'Planting date and nitrogen effects on Fusarium head blight and leaf spotting diseases in spring wheat', *Agron. J.*, 99(1): 113–21.

Summerell, B. A. and Burgess, L. W. (1989), 'Factors influencing survival of *Pyrenophora tritici-repentis*: stubble management', *Mycol. Res.*, 93(1): 38–40.

Summerell, B. A., Klein, T. A. and Burgess, L. W. (1988), 'Influence of stubble management practices on yellow spot of wheat', *Plant Prot. Q.*, 3(1): 12–13.

© Burleigh Dodds Science Publishing Limited, 2017. All rights reserved.

Taylor, R. G., Jackson, T. L., Powelson, R. L. and Christensen, N. W. (1983), 'Chloride, nitrogen form, lime, and planting date effects on take-all root rot of winter wheat', *Plant Dis.*, 67(10): 1116–20.

Unger, P. W. (1994), 'Residue management – what does the future hold?', pages 425–32, in *Managing Agricultural Residues*, P. W. Unger (Ed.), Lewis Publishers, Boca Raton.

Vamshidhar, P., Herrman, T. J., Bockus, W. W. and Loughin, T. M. (1998), 'Quality response of twelve hard red winter wheat cultivars to foliar disease across four locations in central Kansas', *Cereal Chem.*, 75(1): 94–9.

Vanterpool, T. C. (1935), 'Studies on browning root rot of cereals: III. Phosphorus-nitrogen relations of infested fields IV. Effects of fertilizer amendments V. Preliminary plant analyses', *Can. J. Res.*, 1935 (13c(4)): 220–50.

van Toor, R. F., Bithell, S. L., Chng, S. F., McKay, A. and Cromey, M. G. (2013), 'Impact of cereal rotation strategies on soil inoculum concentrations and wheat take-all', *N.Z. Plant Protect.*, 66: 204–13.

Wachowska, U. and Glowacka, K. (2014), 'Antagonistic interactions between *Aureobasidium pullulans* and *Fusarium culmorum*, a fungal pathogen of wheat', *BioControl*, 59(5): 635–45.

Watkins, J. E. and Boosalis, M. G. (1994), 'Plant disease incidence as influenced by conservation tillage systems', pages 261–283, in *Managing Agricultural Residues*, P. W. Unger (Ed.), Lewis Publishers, Boca Raton.

Wegulo, S. N. (2012), 'Factors influencing deoxynivalenol accumulation in small grain cereals', *Toxins*, 4(11): 1157–80.

Wegulo, S. N., Baenziger, P. S., Hernandez Nopsa, J., Bockus, W. W. and Hallen-Adams, H. (2015), 'Management of Fusarium head blight of wheat and barley', *Crop Prot.*, 73: 100–7.

Wegulo, S. N., Bockus, W. W., Hernandez Nopsa, J., De Wolf, E. D., Eskridge, K. M., Peiris, K. H. S. and Dowell, F. E. (2011a), 'Effects of integrating cultivar resistance and fungicide application on Fusarium head blight and deoxynivalenol in winter wheat', *Plant Dis.*, 95(5): 554–60.

Wegulo, S. N., Breathnach, J. A. and Baenziger, P. S. (2009), 'Effect of growth stage on the relationship between tan spot and spot blotch severity and yield in winter wheat', *Crop Prot.*, 28(8): 696–702.

Wegulo, S. N., Hein, G. L., Klein, R. N. and French, R. C. (2008), Managing wheat streak mosaic, University of Nebraska-Lincoln, Extension Circular (EC1871).

Wegulo, S., Stevens, J., Zwingman, M. and Baenziger, P. S. (2011b), 'Yield response to fungicide application in winter wheat', pages 227–44, in *Fungicides for Crop Protection*, InTech, Rijeka.

Wegulo, S., Zwingman, M., Breathnach, J. A. and Baenziger, P. S. (2011c), 'Economic returns from fungicide application to control foliar fungal diseases in winter wheat', *Crop Prot.*, 30(6): 685–92.

Weller, D. M. and Cook, R. J. (1983), 'Suppression of take-all of wheat by seed treatments with fluorescent pseudomonads', *Phytopathology*, 73(3): 463–9.

Wiese, M. V. and Ravensncroft, A. V. (1975), '*Cephalosporium gramineum* populations in soil under wheat cultivation', *Phytopathology*, 65(10): 1129–33.

Wiik, L. and Rosenqvist, H. (2010), 'The economics of fungicide use in winter wheat in southern Sweden', *Crop Prot.*, 29(1): 11–19.

Willyerd, K. T., Li, C., Madden, L. V., Bradley, C. A., Bergstrom, G. C., Sweets, L. E., McMullen, M., Ransom, J. K., Grybauskas, A., Osborne, L., Wegulo, S. N., Hershman, D. E., Wise, K., Bockus, W. W., Groth, D., Dill-Macky, R., Milus, E., Esker, P. D., Waxman, K. D., Adee, E. A., Ebelhar, S. E., Young, B. G. and Paul, P. A. (2012), 'Efficacy and stability of integrating fungicide and cultivar resistance to manage Fusarium head blight and deoxynivalenol in wheat', *Plant Dis.*, 96(7): 957–67.

Wolfe, M. S. (1985), 'The current status and prospects of multiline cultivars and variety mixtures for disease resistance', *Annu. Rev. Phytopathol.*, 23: 251–73.

Xue, A. G., Chen, Y., Sant'anna, S. M. R., Voldeng, H. D., Fedak, G., Savard, M. E., Längle, T., Zhang, J. X. and Harman, G.E. (2014a), 'Efficacy of CLO-1 biofungicide in suppressing perithecial production by *Gibberella zeae* on crop residues', *Can. J. Plant Pathol.*, 36(2): 161–9.

Xue, A. G., Chen, Y. H., Voldeng, H. D., Fedak, G., Savard, M.E., Längle, T., Zhang, J. and Harman, G. E. (2014b), 'Concentration and cultivar effects on efficacy of CLO-1 biofungicide in controlling Fusarium head blight of wheat', *Biol. Control*, 73: 2–7.

© Burleigh Dodds Science Publishing Limited, 2017. All rights reserved.

Xue, A. G., Tenuta, A., Tian, X. L., Sparry, E., Etienne, M., Gaudet, D. A., Menzies, J. G., Graf, R. J., Falk, D. E., Smid, A. and Pandeya, R. S. (2007), 'Evaluation of winter wheat genotypes and seed treatments for control of dwarf bunt in Ontario', *Can. J. Plant Pathol.*, 29(3): 243–50.

Yang, M.-M., Mavrodi, D. V., Mavrodi, O. V., Bonsall, R. F., Parejko, J. A., Paulitz, T. C., Thomashow, L. S., Yang, H.-T., Weller, D. M. and Guo, J.-H. (2011), 'Biological control of take-all by fluorescent *Pseudomonas* spp. from Chinese wheat fields', *Phytopathology*, 101(12): 1481–91.

Yang, M.-M., Wen, S.-S., Mavrodi, D. V., Mavrodi, O. V., von Wettstein, D., Thomashow, L. S., Guo, J.-H. and Weller, D. M. (2014), 'Biological control of wheat root diseases by the CLP-producing strain *Pseudomonas fluorescens* HC1-07', *Phytopathology*, 104(3): 248–56.

Zadoks, J. C., Chang, T. T. and Konzak, C. F. (1974), 'A decimal code for the growth stages of cereals', *Weed Res.*, 14(6): 415–21.

Zhao, Y., Selvaraj, J. N., Xing, F., Zhou, L., Wang, Y., Song, H., Tan, X., Sun, L., Sangare, L., Folly, Y. M. E. and Liu, Y. (2014), 'Antagonistic action of *Bacillus subtilis* strain SG6 on *Fusarium graminearum*', *PloSONE* 9, e92486 doi:10.1371/journal.pone.0092486.

© Burleigh Dodds Science Publishing Limited, 2017. All rights reserved.

Chapter 3

Integrated disease management of barley

Adrian C. Newton, James Hutton Institute and SRUC, UK; and Henry E. Creissen, Neil D. Havis, and Fiona J. Burnett, SRUC, UK

1 Introduction

It is widely recommended and required by EU Directive law (2009/128/EC, article 14) that member states should promote low pesticide input pest management; in effect that all farmers and growers have an integrated pest management (IPM) plan. This is no less important for broad acre cereal crops like barley than it is for, for example, tomato crops in glasshouses. However, the perception is that the concept applies more in the latter type of crop as it requires the use of carefully managed biocontrol agents to eliminate fungicides and their residues from the system. Whilst it may be true that the latter represents very successful application of IPM principles, the same principles can and should be applied to the former, and many growers do to some degree. Nevertheless, in many countries most farmers have no IPM plan as such (76% in Scotland: Scottish Government, 2017) but to effectively manage diseases in barley, this must be carried out in an IPM context to be sustainable.

http://dx.doi.org/10.19103/AS.2019.0060.18
© Burleigh Dodds Science Publishing Limited, 2020. All rights reserved.

A definition of IPM is 'the careful consideration of all available pest control techniques and subsequent integration of appropriate measures that discourage the development of pest populations and keep pesticides and other interventions to levels that are economically justified and reduce or minimize risks to human health and the environment. IPM emphasizes the growth of a healthy crop with the least possible disruption to agro-ecosystems and encourages natural pest control mechanisms'; or more succinctly: 'Integrated Pest Management (IPM) is an ecosystem approach to crop production and protection that combines different management strategies and practices to grow healthy crops and minimize the use of pesticides' (FAO, 2018).

How do these principles apply to disease management in barley production? We will describe the main options available in the sections below and consider them in the key context of integration. They may also be considered in the context of management decisions under the headings of (1) seed-based, (2) mechanical/growing medium, (3) cropping/biological control, (4) timing and (5) input efficiency (Table 1). Integration can mean simply additive effects of the components, but more than likely is that some components will interact synergistically while others may be antagonistic. Furthermore, the outcomes will be different whether considered from an economic, ecological or sociological context. However, in many cases we do not know how they interact in any of these contexts and research is needed to determine the mechanisms and develop tools to aid their integration in practice.

IPM can include all possible biotic and abiotic stress factors or influences but as the title implies, we focus on disease and generally exclude for example, weeds, nematodes and insects as these usually require different approaches and are therefore best considered separately. Furthermore, diseases represent the biggest threats to barley yield for much of the barley growing area not already constrained by abiotic stress, and for which there are many management options.

2 Barley production context: requirements and constraints

Barley is grown for animal feed, human food, as a substrate for grain or malt distilling or for brewing, and as whole crop for feed, silage or anaerobic digestion (Newton et al., 2011). For each of these uses different cultivars, mixtures or even populations will be grown in very different environments with different agronomies. Management and inputs can be from minimal to intensive and therefore outcomes in economic, ecological or sociological contexts may be very contrasting. For example, a malting barley crop for the Scotch whisky industry must be of a single, pure, often-specified cultivar and achieve specific grain size, grain nitrogen and germination standards. In contrast, a whole crop of barley for silage may need to produce the highest quantity of digestible dry matter from negligible inputs. The consequences for pests and pathogens as well as the options available will be very different. In the former case herbicide and fungicide inputs will be required and there is scope for minimising them with monitoring, forecasting and interpretation of crop intelligence data. In the latter case herbicides and fungicides may not be required because the risks and consequences of not applying them are low due to the inherent qualities of the system. However, there may be other options that cannot be considered in the malting crop such as companion- or inter-cropping.

© Burleigh Dodds Science Publishing Limited, 2020. All rights reserved.

Table 1 IPM categories for consideration in an ecological, economic or sociological context

1) *Seed based*
Using certified seed
Selecting cultivars with resistance to key risks in your area
Testing home-saved seed
Use appropriate seed treatments on affected or early drilled crops
Stale seedbeds
Cultivar diversification
Use of cultivar mixtures where markets allow
Alteration of crop/plant population density
2) *Mechanical/growing medium*
Direct drilling
Minimum tillage
Regular ploughing
Rotational ploughing
Under-sowing of crops
Increasing soil organic matter
Choice of cover crops
3) *Cropping/biological control*
Summer fallow
Encourage beneficial organisms
Use of biological controls
Catch/cover crops to manage soil pests (e.g. PCN)
4) *Timing*
Planning crop nutrition
Planning drilling date
Planning drainage requirements
An agri-environment scheme requirements
5) *Input efficiency*
Use of treated seed
Spot treatment
Drift reduction technology
Precision application technology
Field mapping
Attention to growth stages, spray timings and intervals
Technical updates on product efficacy
Crops monitored pre- and post-treatment
Thresholds, risk warnings or weather data used
Where feasible to do so reduce pesticide dosage and/or frequency of pesticide use
Incorporate resistance elicitor-based products in crop protection programme where available
Calibration and maintenance of sprayers
Pesticide resistance stewardship advice followed

© Burleigh Dodds Science Publishing Limited, 2020. All rights reserved.

Winter and spring barley systems are another contrast where the presence or absence of a cold period required for vernalisation also has consequences for pests and pathogens. However, other barley crops are winter sown to better utilise rain and some crops are grazed before then going on to grain harvest. Such contrasting systems cannot be considered for all the components of IPM so below we will consider the options available in the systems where they have potential to contribute to beneficial outcomes.

3 Diseases overview

Barley is plagued by a great diversity of diseases caused by a wide range of fungal, viral and bacterial pathogens, many of which can seriously impact upon the crop's ability to produce high yields and quality grains. When considering the management of barley disease it is important to understand which diseases pose the greatest threat and the management strategies that are effective in controlling them. Due to differences in life cycles it may be that different diseases require different management strategies which may sometimes require conflicting approaches, for example reduced tillage regime to control common eyespot (Burnett et al., 2012) or ploughing to control tan spot (Bockus and Claassen, 1992). Table 2 summarises information on many of the current economically important barley diseases, their causal agents, suitable management strategies and issues to consider when attempting to control them. It's important to understand that over-reliance on a single control measure puts a selection pressure on the pathogen to evolve and overcome the control measure, thus an integrated approach to disease management is required if the disease is to be managed in a sustainable fashion.

4 Inoculum management: sources and epidemiological conditions

Two major considerations determine whether a disease has the potential to become problematic: (1) whether the inoculum is present and (2) whether favourable epidemiological conditions exist, for both host and environment. In addition, not all pathogens cause enough yield or quality losses to be economically viable to treat even in their most severe cases. IPM choices should be determined by this knowledge. For inoculum that is brought predominantly by wind such as the rusts and mildews, crop intelligence reports can be very useful for determining risk and therefore choice of appropriate treatments. Knowledge of outbreak locations, virulence specificity of the inoculum and fungicide resistance traits can all be used to aid decisions, although some of these data will be interpolated from trends determined by full characterisation of previous years' isolates. Examples of this include the UK Cereal Pathogen Virulence Survey (UKCPVS, 2018) where the previous year's data is disseminated before the growing season starts and updates about outbreaks are released during the season for powdery mildew and yellow rust.

A second source of incoming inoculum is seed-borne. For barley pathogens this largely affects rhynchosporium (barley leaf blotch or scald) and ramularia (Ramularia leaf spot), though it can be important for net blotch too. Here clean seed or adequate seed

© Burleigh Dodds Science Publishing Limited, 2020. All rights reserved.

Table 2 Summary of the most economically important barley diseases, the biology of their causal agents and potential control methods

Disease	Pathogen	Biotroph/ hemibiotroph/ necrotroph	Target (foliar/ ear/stem base/root)	Dispersal method	Occurrence	Economic importance	Genetic resistance (mg, pr)	Pesticide resistance $(1 \rightarrow 3)$ low→high)	Cultural control	References
Fungi										
Powdery mildew	*Blumeria graminis* f. sp. *hordei*	Biotroph	Foliar	Wind	Very common	Medium	mg, pr	2	Control volunteers, crop residue removal, don't overfertilise	Lyngkjaer et al. (2000), Kusch and Panstruga (2017)
Brown rust	*Puccinia hordei*	Biotroph	Foliar	Wind	Common	Medium	mg, pr	1	Control volunteers	Grasso et al. (2006)
Stem rust	*Puccinia graminis* f. sp. *secalis*, *Puccinia graminis* f. sp. *tritici*	Biotroph	Foliar	Wind	Very rare	High	mg, pr	1	Control volunteers	Grasso et al. (2006), Kleinhofs et al. (2009)
Yellow/ stripe rust	*Puccinia striiformis* f. sp. *hordei*	Biotroph	Foliar	Wind	Rare	Medium	mg, pr	1	Control volunteers	Chen et al. (1994), Yan and Chen 2006
Leaf scald	*Rhynchosporium commune*	Hemibiotroph	Foliar	Seed, trash	Very common	High	mg, pr	2	Rotation, crop residue removal, delay sowing	Kendall et al. (1994), Oxley and Burnett (2009), Zhan et al. (2008), Cooke et al. (2004)
Ramularia leaf spot	*Ramularia collo-cygni*	Hemibiotroph	Foliar	Seed, wind	Common	Medium	Unidentified	3	None known	Walters et al. (2008), Havis et al. (2015)
Net blotch (net form)	*Pyrenophora teres* f. *teres* (net form)	Necrotroph	Foliar	Trash, seed	Rare	High	mg, pr	3	Crop rotation, crop residue removal	Semar et al. (2007)
Net blotch (spot form)	*Pyrenophora teres* f. *maculata* (spot form)	Necrotroph	Foliar	Trash, seed	Rare	Medium	mg, pr	3	Crop rotation, crop residue removal	Semar et al. (2007)

(Continued)

© Burleigh Dodds Science Publishing Limited, 2020. All rights reserved.

Table 2 (*Continued*)

Disease	Pathogen	Biotroph/ hemibiotroph/ necrotroph	Target (foliar/ ear/stem base/root)	Dispersal method	Occurrence	Economic importance	Genetic resistance (mg, pr)	Pesticide resistance (1→3 low→high)	Cultural control	References
Tan spot	*Pyrenophora tritici-repentis*	Necrotroph	Foliar	Trash, seed	Rare	Medium	mg, pr	NA	Crop rotation, crop residue removal	Bockus and Claassen (1992), Carignano et al. (2008), Duczek et al. (1999)
Stagonospora blotch	*Parastagonospora nodorum*	Hemibiotroph	Foliar	Trash	Rare	Low	mg, pr	2	Crop rotation, crop residue removal	Duczek et al. (1999), Caten and Newton (2000)
Head/ear/ seedling blight	*Microdochium nivale, Fusarium graminearum* but also *F. avenaceum, F. culmorum, F. poae*	Necrotroph	Ear/stem base	Soil	Common	Medium	pr	2	Plough, rotation not including maize	Bateman et al. (2007), Cromey et al. (2006), Browne and Cooke (2005), Bai and Shaner (2004), Scherm et al. (2013)
Ergot	*Claviceps purpurea*	Necrotroph	Ear	Seed	Very rare	Low	Unidentified	3	Rotation, plough, weed control	Orlando et al. (2017)
Cephalosporium stripe	*Cephalosporium gramineum*	Necrotroph	Foliar	Trash/soil	Rare	Low	Unidentified	1	Rotation, plough, weed control	Havis and Gorniak (2004)
Take all	*Gaeumannomyces graminis var tritici*	Necrotroph	Root	Soil	Rare	Low	Unidentified	NA	Rotation, delay sowing	Oxley and Burnett (2009), Cromey et al. (2006)
Common eyespot	*Oculimacula yallundae, O. acuformis*	Necrotroph	Stem base	Trash, soil	Rare	Low	mg, pr	2	Control volunteers, rotation, delay sowing	Burnett and Hughes (2004), Burnett et al. (2012)
Sharp eyespot	*Ceratobasidium cereale*	Necrotroph	Stem	Trash, soil	Rare	Low	pr	2	Control volunteers, control weeds, plough	Colbach et al. (1997), Cromey et al. (2006)

© Burleigh Dodds Science Publishing Limited, 2020. All rights reserved.

Disease	Pathogen	Trophism	Site	Transmission	Frequency		Resistance	No.	Control	Reference
Loose smut	Ustilago nuda f. sp. hordei	Biotroph	Ear	Seed	Very rare	Low	pr	2	Clean seed	Menzies et al. (2009), Menzies et al. (2010)
Covered smut	Ustilago hordei	Biotroph	Ear	Seed	Very rare	Low	pr	1	Clean seed	Menzies et al. (2009)
Leaf stripe	Pyrenophora graminea	Necrotroph	Foliar	Seed	Rare	Low	pr	1	Rotation	Pecchioni et al. (1999)
Bacteria										
Bacterial leaf streak	Xanthomonas campestris pv. hordei	Necrotroph	Foliar	Seed, splash	Very rare	Low	pr	NA	Plough, remove weeds	Alizadeh et al. (1994), El Attari et al. (1998)
Bacterial leaf blight /basal kernel blight	Pseudomonas syringae pv. syringae	Necrotroph	Foliar	Wind, splash, seed	Very rare	Low	pr	NA	Plough, clean seed	Braun-Kiewnick et al. (2000)
Viral										
Barley yellow mosaic disease	Barley yellow mosaic virus (BaYMV), Barley mild mosaic virus (BaMMV)	Biotroph	Foliar	Soil (Polymyxa graminis)	Rare	Low	Unidentified	NA	None known	
Barley yellow dwarf virus (BYDV)	Barley yellow dwarf virus (BYDV)	Biotroph	Foliar	Aphid vector	Common	High	Unidentified	NA with respect to virus. 3 with respect to aphid vector	Ploughing/destroying crop residues several weeks before drilling, rotation, altering sowing dates to limit contact with vector	Miller and Rasochova (1997)

mg = major gene, pr = partial resistance.
NA = Not applicable as pesticides rarely used.

© Burleigh Dodds Science Publishing Limited, 2020. All rights reserved.

treatment with fungicide, heat, resistance elicitor or biocontrol agent may reduce or eliminate the inoculum source. It can also control other inoculum such as the smuts and ergot. Inoculum that is predominantly on the seed surface is effectively treated in this way but ramularia inoculum is often present throughout the seed including the embryo, so effective treatment is very difficult without risking seed viability or vigour loss (Havis et al., 2014b). Indeed, ramularia colonises the whole plant endophytically, and we will consider the implications of endophytes in more detail later.

As pathogens that are seed-borne are generally non-obligate fungi they are usually already present in a third source, the crop environment, surviving on crop debris or soil depending on their saprophytic growth characteristics. Therefore, agronomic factors and general phytosanitation factors can be manipulated to reduce risk. Soil cultivation, particularly inversion tillage (ploughing) where the plant debris is buried, compared with non-inversion tillage (minimum tillage or direct drilling) can make a difference particularly to early season disease.

A fourth source is volunteer plants and reservoirs in adjacent crops or crop margins. This is a potential increasing problem for non-specialised inoculum such as ergot, but where host specificity even at a species level is important, such as for rhynchosporium (King et al., 2013; Penselin et al., 2016), inoculum and symptoms on neighbouring plants are low risk. The risk from volunteer and other reservoir plants as hosts to any type of pathogen can be reduced only by appropriate agronomic decisions and phytosanitary measures. This might include rotation, probably the most effective tool for rhizoctonia and take-all control.

Another source of disease that could be from any of the original inoculum sources is asymptomatic infection, that is they are not causing disease but infection can be extensive. Such infected plants require signals that cause the transition to pathogenicity. Such signals could be developmental, abiotic stress or temporal (Newton et al., 2010). However, in an IPM context we should consider host–microbe interactions in a broader context by managing our crops in such a way that benign and beneficial associations are favoured and that pathogenic or parasitic associations are disadvantaged (Looseley and Newton, 2014). An example of the former would be to encourage mycorrhizal associations which can be achieved by minimising soil disturbance in our agronomic management. Knowledge of the source of inoculum can help to target treatment too, that is whether seed-borne, soil, trash, foliar for wind or splash-dispersed transmission.

Having considered (1) whether inoculum is present, if it is then we must consider (2) whether favourable epidemiological conditions exist. A good example of this is the effect of seed heavily infected with rhynchosporium versus clean seed on epidemic development. Here it was found that in some seasons the seed-borne inoculum had a big impact on epidemic development, but in others it was the later epidemiological conditions that were dominant and the extra seed-borne inoculum had no impact (Havis et al., 2014a). The general nature of the favourable epidemiological conditions can be deduced for each disease. These include wet conditions, particularly heavy rain for splash-dispersed pathogens such as rhynchosporium, especially during stem extension which promotes transmission up the canopy. Other conducive conditions include dew on the leaves during cool nights, which favours rust infection; and warm, humid conditions without free water which promotes mildew infection. However, for some pathogens such as ramularia and rhynchosporium there are also stress factors that are less clearly understood making epidemic prediction less certain (Newton et al., 2010).

© Burleigh Dodds Science Publishing Limited, 2020. All rights reserved.

5 Varietal resistance

Use of durable resistance is the obvious and most effective strategy to deploy in IPM. However, durable resistance in practice may take many forms and should be a product of its context, that is considering all the environment factors too, both natural and managed. There are very few examples of single major genes that have proved durable and the general strategy should be to not use them as the main component of a resistance strategy. To do so would be to expose them to such strong selection that the pathogen will mutate to virulence rendering the resistance ineffective. The exceptions would be where other components of the IPM strategy are more effective resulting in very little selection pressure, or an unusual type of genetic resistance such as the *mlo* gene in barley that has remained durable despite widespread use in spring barley (Kusch and Panstruga, 2017). However, the more likely genetic basis of durable resistance is variously referred to as polygenic, partial resistance or field resistance. Crucially, it comprises multiple expressions of resistance mechanisms that together amount to effective, although rarely total, resistance. The components separately are likely expressions of different genes each of which has a small effect on resistance, the net effect of which is additive or even synergistic. On resistance ratings such as one (susceptible) to nine (fully resistant), AHDB Recommended lists (AHDB, 2018) these are likely to be seven or eight but rarely nine.

Partial resistance is not necessarily either polygenic or race-non-specific and there are examples of where it is neither (Newton, 1989). Therefore, not relying on a single genotype whatever its resistance phenotype is a good strategy for reducing risk. Expressed another way, diversification is key to durable or effective resistance in practice. This can be from crop types through region and fields to within a crop or even as intercrops. The deployment of the barley *mlo* gene is an example of crop diversification between crop types as it is deployed in spring barley but not in winter barley at present (Lyngkjær et al., 2000). Diversification schemes have been published in the past to encourage farmers to grow cultivars with different resistance genes in neighbouring fields. These schemes are devised based on knowledge of the virulence frequencies or race composition of the current pathogen population (UKCPVS, 2018). Within-crop diversity is effective in cultivar mixtures or blends, or more rarely where multilines have been developed, that is a common cultivar with different resistance genes back-crossed into it (Mundt, 2002). Again, ideally these need to be designed with respect to the composition of the pathogen populations that affect the crop, and complex strategies can be very effective (Newton and Guy, 2011). Mixtures must be designed with all other aspects of the interaction for the end use in mind too (Newton et al., 2009). However, multiple components and patchiness are the key to enhancing the effectiveness of these strategies (Newton et al., 1997; Newton and Guy, 2009). Intercrops can be regarded as like cultivar mixtures but where the trait differences are much larger. Nevertheless, the principles of more diversity and heterogeneity prevail (Newton et al., 2009).

Disease tolerance is a concept little exploited in IPM as it is hard to define (Newton, 2016). A simple definition is less yield loss than might be expected from the disease severity observed but this begs many questions about disease assessment and the disease–yield relationship of crops with respect to resources (Bingham and Newton, 2009; Bingham and Topp, 2009; Bingham et al., 2009), that is resource use efficiency. However, cultivar mixtures too have complex relationships with resource use efficiency, disease and thereby tolerance and in all these considerations we find that it is a balance between the

© Burleigh Dodds Science Publishing Limited, 2020. All rights reserved.

negative effects of competition and the positive effects of facilitation between plants, be it in monocultures or mixtures (Brooker et al., 2016).

Disease escape is a simpler concept to understand and potentially to exploit as it is mechanistically based. In splash-dispersed pathogens height is a major component. A mapping population of barley segregating for two dwarfing genes shows two strong quantitative trait loci (QTL) for rhynchosporium resistance. These are in fact the loci of the dwarfing genes as shorter plants get more splash onto the upper foliage that then infect and cause disease (Looseley et al., 2012). Erectoid leaf habit also reduces infection as fewer leaves are exposed to splash. Other leaf habit attributes affect the microclimate of the canopy and thus affect likelihood of successful infection. All of these are escape mechanisms rather than resistance but can be deployed in a complementary way.

Breeders exploit molecular marker technology increasingly to enhance the likelihood of desirable traits being expressed in advanced selections under field conditions. However, these are used effectively only on genes with strong expression and many effective resistance expressions are the product of many genes each of small effect. Some virus resistance such as Barley Mild Mosaic Virus and Barley Yellow Mosaic Virus are routinely screened for in breeding. However, among the fungal diseases only the *mlo* gene for powdery mildew resistance is both oligogenic, strongly expressed and durable, although other genes with useful levels of resistance such as alleles *Rrs1* for resistance to rhynchosporium could be selected. While genetic modification might appear to offer a potential fast-track to resistance deployment in some crops, in barley not only are there so few candidates, but also most varietal backgrounds with the exception of Golden Promise have proved to be intransigent. Gene editing is likely to offer considerably enhanced opportunities, particularly with the advent of the barley Pan-Genome projects such as SHAPE at IPK Gatersleben in Germany which aims to unravel the complexity of the core genome shared in all barley haplotypes and the 'dispensable' part of the genome found only in individual haplotypes. Such technologies with these bioinformatic resources could lead to strategies for pyramiding or cassettes of resistance components that should be durable and therefore very valuable. However, commercial varieties exploiting these resources still have to pass DUS (Distinctiveness, Uniformity and Stability) and VCU (Value for Cultivation and Use) regulations in most markets for widespread availability and exploitation, and must still be managed as components of IPM to maximise their effectiveness and maintain their durability.

6 Crop protectants

One of the core principles of IPM systems is to promote crop management strategies which reduce reliance on pesticides (Lamichhane et al., 2016). The application of crop protectant agents in targeted and sustainable ways is therefore a component of all IPM systems. In practical terms most crop protectant agents applied to barley are fungistatic and classed as pesticides, although alternatives such as elicitors, biologicals and biostimulants are also the subject of research effort, with some very limited uptake in practice (Chandler et al., 2011). The use of pesticides carries cost implications as well as benefits and can be contentious (Geiger et al., 2010). They are heavily regulated, both in terms of their inherent safety/toxicity to operators, consumers and the wider environment and also in terms of the dosage, timings, skills and equipment required. In addition, specific criteria

© Burleigh Dodds Science Publishing Limited, 2020. All rights reserved.

need to be met to satisfy regulators that efficacy claims are justified and to demonstrate that the crop protection agents are not phytotoxic on the target crop (Regulation (EC) No 1107/2009 of the European Parliament). There are many practical considerations to be overcome too, such as ease of manufacture and stability and shelf life.

Fungicides are defined as chemical agents that kill a fungal pathogen or limit its growth and they can be synthetic or natural compounds. Their mode of action (MOA) is not always known or understood prior to launch but usually they are effective by interfering with key metabolic process or pathways within the target pathogen and can be grouped accordingly by their MOA. Knowledge of their MOA is useful for inferring their likely field of efficacy, for example strobilurin fungicides which target energy production are far more effective in inhibiting spore germination than in preventing later mycelial growth (Hamdy, 2007). MOA is also useful in terms of developing stewardship and anti-resistance strategies which aim to incorporate a diversity of modes of actions in control programmes, and also to protect non-target species such as aquatic life and bees.

Other crop protectant agents may not be fungistatic as such but can be effective by switching on the host plant's own defence mechanisms so that it is better able to detect and recognise a pathogen early in the infection process (Walters et al., 2014). These are often termed resistance elicitors. Biological control agents are more common in terms of control of insect pests than they are for fungal pests, for example nematodes that kill leatherjackets (Kergunteuil et al., 2016), although these are costly and only used in high-value contexts such as sports turf and airports and not in the arable context to protect barley.

Crop protection agents can have both fungistatic effects and also effects on the host plant, which may be as obvious as the growth regulatory or greening effects seen with some fungicide groups. Biostimulant is a term used to cover products applied to the crop which have greening or nutritional benefit, some of these have potential to allow a crop to yield well even in the presence of disease, and some biostimulants may also elicit host defence mechanisms (Xu et al., 2011).

Key principles in the integrated use of fungicides in managing a barley disease is that their use is targeted, not just at the key pathogen risks but also targeted in terms of timings (Jørgensen et al., 2017). Control programmes should be based on the identification of the key disease targets, but also predicated on the key timings of application and what their purpose is. Barley is generally recognised as being sink limited – in other words that yield potential is more driven by the number of grain sites than it is by the source, that is the amount of photosynthates captured by green leaf (Kennedy et al., 2017). This means that retaining tillers, and hence grain sites, is crucial early in crop development and why late tillering/start of stem extension is a key timing for maximising yield response to fungicides in both the winter and spring crop. Prior to this in an IPM system any earlier disease would only be targeted if indicated by risk – for example the very early ingress of mildew, brown rust or other foliar diseases in the autumn or early spring. Later sprays to retain green leaf would be used to manage established disease not well controlled by this first timing and to retain green leaf, which is particularly important in the spring sown crop. The spring crop has less capacity to compensate for periods of poor growth so protection of later green leaf is often required where the site or cultivar need dictates. Protection against ramularia is an example of a late season treatment often applied to barley crops and timing is dictated by the need to protect crops prior to the stress of flowering, and also by last application dates for many fungicides. Fungicide applied after ear emergence gives

© Burleigh Dodds Science Publishing Limited, 2020. All rights reserved.

an example of a timing where yield responses are very small and seldom exceed the cost of application, and which in an integrated management programme are easily dropped.

Crop protection programmes should be reactive to risk but largely built around these key crop timings where benefits to yield are maximised. The fungicide choice, dose rate and mixing partners should target the main disease threats for the barley crop in question. The key disease risks for each crop should be assessed prior to designing crop protection programmes and then modified in response to emerging in-season intelligence as described above (Hughes et al., 1999). This can be formal knowledge of risk systems and crop monitoring, but also informal knowledge of the endemic disease risks in an area – for example rhynchosporium may be an annual threat in wet areas where there is a history of growing susceptible cultivars. Programmes should be tailored to the risk in individual fields, set within the context of a sustainable farm system and rotation. The growth of continuous spring barley has been discouraged through farmer direct support payments which incentivise diverse rotations by making payments conditional on cross-compliance with good practice (Council Regulation 1306/2013), but is an example of poor practice which can build-up trash-borne issue such as Cephalosporium and stem base diseases like eyespot. The targeting of inputs can extend as far as precision surveying and application techniques allow – so spot treatment or applications to areas of fields identified as at higher risk can be part of an IPM system (O'Grady and O'Hare, 2017). Spot liming barley crops is already practised and as technological solutions present themselves this could encompass remote detection and application to specific areas of fields. Early identification of risk allows for pre-planting control measures to be implemented such as a different crop choice or the use of a cultivar with resistance to the main threat.

6.1 Fungicide methods of application

Fungicides can be applied as seed treatments, foliar sprays or soil treatments (Matthews, 2000). The latter tend to have high costs and significant negative environmental impacts and so there are no current examples used in barley. Seed treatments are a targeted and effective way of addressing seed-borne problems. Seed treatments can also have some effect in reducing trash and soil-borne diseases such as take-all. Seed treatments are an effective means of managing seed-borne barley health issues such as loose smut, leaf stripe and seed-borne net blotch. They can also help to manage Fusarium and Microdochium species in the context in which these pathogens can lead to seedling blights. They should always be used in conjunction with seed testing and with the purchase and drilling of seed of a high health status, for example as purchased through official certification schemes. Seed treatments should be used, as with other fungicide inputs, as part of an integrated package and should not be used to pull seed with known health issues up to acceptable levels of seed health. Some seed treatments will also move into emerging leaf tissue and protect against early season foliar diseases so used to be targeted at early sown and high-risk crops. However, as these foliar-acting seed treatments tend to be applied at higher doses they present a greater potential hazard and so have been removed from the European market, leaving what are often termed the 'single-purpose' seed treatments, that is those purely targeting the classic seed-borne health issues. The addition of silthiofam (to manage take-all) is an example of a seed treatment that can be added to extend the range of the 'single-purpose' treatments where take-all is deemed a risk. Other measures to reduce take-all such as extended rotation and late drilling should of course be practised as a first resort.

© Burleigh Dodds Science Publishing Limited, 2020. All rights reserved.

Fungicides are also applied as foliar treatments to protect against foliar pathogens. Fungicides are usually applied to barley crops in programmes which aim to protect the crop over the growing season and, as indicated above, should be targeted with the key timings and the key disease risk in mind (Young et al., 2006).

6.2 Fungicide modes of action

Knowledge of the characteristic of each fungicide is useful in targeting their use to the situations where they will be most effective. Some fungicides are termed protectant and will act early in fungal life cycles, such as preventing spore germination, but will be less effective as eradicants where disease is established. Some fungicides are more eradicant in nature and can be used to target established infections – for example the morpholines. Many fungicide groups sit somewhere on this continuum – for example the demethylation inhibitors (DMIs) or succinate dehydrogenase inhibitors (SDHIs) are fungicide groups both containing fungicide actives that give both eradicant or protectant activity (Oliver and Hewitt, 2014). There is a potential conflict in IPM in terms of applying fungicides prophylactically and getting maximum efficacy using lower doses or waiting until thresholds are reached and disease is established, and using fungicides eradicantly (also termed curatively) but at higher doses. This is why as broad a consideration of risk as possible is helpful in identifying historic and regional risk and allowing key risks to be targeted in a timely manner. Using a diversity of fungicide MOAs allows for a range of crop growth stages and diseases to be targeted in a crop and for varied programmes to be developed and tailored to the crops in question. In addition, diversifying MOAs in barley programmes reduces reliance on any one component with potential benefits to the environment and with benefits to fungicide resistance management strategies (FRAC Code List, 2018). MOAs with activity against barley pathogens include quinone outside inhibitors (QoIs), anilino-pyrimidines, methyl benzimidazole carbamates (MBCs), morpholines, a phthalonitrile and dithio-carbamates, as well as the DMI or SDHI fungicide groups mentioned above.

6.3 Fungicide stewardship

Although there are strict testing regimes before pesticides are registered for use and statutory limitations on their usage, the use of fungicides can still be contentious, and the health and environmental impacts are kept under constant review (European Union, 2009). Many pathways are shared between fungal pathogens and other organisms so fungicide groups under review currently include the DMI group of actives where endocrine disruption is a health concern for some actives within this group, or chlorothalonil which has been linked to bee declines in the United States (McArt et al., 2017) which breaks down into multiple metabolites, not all of which can be assessed accurately. Statutory limitations such as buffer strips around fields and restrictions around water courses, maximum dosages, harvest intervals and other restrictions on use are commonly applied. Adherence is a legal requirement.

Some countries, such as Denmark have further disincentivised fungicide use by imposing measures such as pesticide taxes. Other countries, such as the United Kingdom have opted for voluntary initiatives (VI) to reduce reliance on pesticides, and market drivers such as supermarket protocols or quality assurance schemes also promote sustainable,

© Burleigh Dodds Science Publishing Limited, 2020. All rights reserved.

responsible and integrated usage. VI measures include such measures as skills and training for sprayer operators, the use of skilled and qualified agronomists (BASIS registered) and other stewardship measures such as testing sprayers, use of low drift nozzles and other methods which reduce the risk drift or damage to non-target organisms.

Fungicide resistance management is a further example of stewardship measures that are applied to barley crops. The selective advantage conferred on resistant individuals in a population, particularly where individual MOAs are intensively used, drives resistance development. Anti-resistance strategies which aim to reduce selection pressure are based on principles of reduced reliance (van den Bosch et al., 2014a). The risk of resistance developing to fungicides that act at a single point in the metabolic chain, termed single-site fungicides, is very high. It is much harder for pathogens to develop resistance against fungicides that act at multiple points in a cell's metabolism and they tend to carry fitness costs for the pathogen and also a higher risk of side effects in non-target species. For this reason, there are few examples of new multi-site actives that have made it through the development and registration processes recently. In barley folpet and chlorothalonil give examples that have been used in the multi-site context, but other multi-sites include older products such as mancozeb and sulphur. To reduce the risk of fungicide resistance, reliance on any single active must be reduced and in practical terms this means incorporating other IPM measures which reduce reliance on chemistry and, when planning a fungicide programme, using a diversity of chemistry. This can be implemented either as mixtures and/or as alternations, and ideally as both. The inclusion of multi-sites is beneficial in reducing risk to the higher risk actives (van den Bosch et al., 2014b). Annual updates and guidance on key crops are provided in the United Kingdom by FRAG-UK and by other groups such as NOROBAG in Nordic countries. FRAC is an agrochemical industry initiative which is pan global, to monitor resistance and formulate strategies.

Barley pathogens have a track record of developing resistance problems with most key pathogens carrying some form of reduced efficacy in the field. Ramularia is a well-documented example with field resistance to all the main single-site fungicides which were previously effective including MBCs, DMIs, SDHIs and QoIs. Other examples of barley pathogens with resistance issues include loose smut (historic examples of SDHI resistance), rhynchosporium (older DMIs), mildew (QoIs, older DMIs), tan spot (SDHIs, QoIs) and net blotch (QoIs and SDHIs). Low doses generally reduce the risk of resistance development (Grimmer et al., 2015). Anti-resistance stewardship is currently a mix of statutory advice on labels (maximum application number) plus 'softer' on label guidance such as the advice to follow FRAG-UK guidance or include a suitable mixing partner, and voluntary guidance such as that advocated through FRAG-UK. Some consideration of resistance risk has to be made prior to registration so that this can be incorporated into statutory controls.

7 Agronomy

Several agronomic components of IPM have already been mentioned, namely reduced cultivation leading to potential enhanced mycorrhizal colonisation but also to enhanced pathogen inoculum, and also rotation length to minimise inoculum, drilling date (spring crop the ultimate example), sowing density, ploughing (increases eyespot) and use of growth

© Burleigh Dodds Science Publishing Limited, 2020. All rights reserved.

regulators. The former is likely to enhance nutrition and prime resistance mechanisms, whereas the latter may enhance disease. Intercropping was mentioned above as a way of exploiting diversity and it will promote facilitation between species if well designed. It can also limit the crop protectants that can be used due to their MOAs, for example herbicides that affect dicots but not monocots, or for regulatory or product approval reasons. Crops can also be sown later before winter to avoid pathogen build-up before the dormant period in colder regions. The crop sequence can be important with respect to soil microbes in general, pathogen inoculum in particular and nutrition and soil structure. An example would be the effect of maize (corn) debris on fusarium inoculum causing problems in subsequent barley and wheat crops. However, here again there are many other factors that affect risk and these can be evaluated in an IPM plan (Edwards, 2013).

Resistance elicitors were referenced above but they can be components of products regulated as biostimulants. Although disease control cannot be claimed as an attribute of biostimulants without them being classified and regulated as pesticides, plant health claims are made and these effects can be of value in IPM protocols. Likewise, many nutritional products that target micronutrients can be beneficial to plant health and resistance expression. However, it is rarely individual nutrients that have an effect irrespective of the status of other major and minor nutrients (Walters and Bingham, 2007) so there are few rules that can be used to optimise the contribution of these. Indeed, the effect of soil pH alone on disease can be considerable as it affects nutrient availability differentially (Holland et al., 2017).

8 IPM knowledge sources and tools

The successful implementation of IPM by growers is dependent on the availability of information on the health status of crops, potential disease threats and changes in pathogen variation. Monitoring of disease is carried out at a regional level in many countries to give an accurate record of annual epidemics and an indication to growers of disease threats. Disease outbreaks are reported via online information platforms such as 'Adopt-a-Crop' in Scotland, 'Crop Monitor' in England and Wales, and Mississippi Crop Situation in the United States (Mississippi State University, 2018). In addition to intelligence on crop disease levels, growers require updates on changes in pathogen variation and fungicide sensitivity. National bodies often undertake monitoring of pathogen variation, for example the National Institute of Agricultural Botany in the United Kingdom monitors changes in the race structure of the yellow rust pathogen of wheat (*Puccinia striiformis*). In the United Kingdom there was a rapid change in race structure of the yellow rust pathogen in 2016 which rendered many previously resistant cultivars sensitive to the disease. Research institutes and agrochemical companies also monitor changes in pathogen sensitivity to fungicides and changes on a larger scale are reported by bodies like the Fungicide Resistance Action Group UK (FRAG-UK). Recent mutations in the barley pathogen *Ramularia collo-cygni* have reduced the efficacy of two major groups of fungicides to control disease symptoms (Havis et al., 2018). This information should be interpreted in the context of on-farm issues such as the historical risk, the recorded effects of diseases on yield and quality and the likely threats locally, all important components of planning IPM strategies on-farm.

© Burleigh Dodds Science Publishing Limited, 2020. All rights reserved.

Many of the principles, concepts and practical information discussed above are implemented in the AHDB Barley Disease Management Guide for the United Kingdom (AHDB, 2016) and many other countries have similar sources of applied information and advice. A risk forecast has been elucidated for a number of barley diseases such as Fusarium head blight and Ramularia leaf spot. However, the ramularia forecast was based on a limited number of environmental parameters and requires more research to become a fully robust forecast (Havis et al., 2018). A risk assessment has been produced for barley yellow dwarf virus based on monitoring of aphid populations (Fabre et al., 2006). The model uses a single assessment of the proportion of plants infested with aphids and then a temperature-dependent simulation of aphid dynamics.

The assessment of risk in any barley crop requires the integration of a number of factors including previous cropping, cultivation methods, crop nutrition, varietal susceptibility, fungicide efficacy and availability of alternative control measures. Each of the major barley pathogens has an optimal temperature for growth and development such as 15–20°C for *Rhynchosporium commune* and 20–25°C for *Pyrenophora teres* (Mathre, 1997). Therefore, the use of weather forecasts combined with actual data should provide growers with an indication of the likelihood of disease development

9 Uptake and communication of IPM

IPM is a knowledge-intensive process in which evidence-supported approaches are selected based on the biotic threats threatening the crop (Byerlee, 1996). It is therefore important to consider the ways in which famers gather and decipher information that ultimately results in change in practice. Farmers constantly receive floods of information and plant protection advice from various sources. Knowledge transfer messages via farming press, advisory body publication, radio broadcasts and so on do not always constitute a consensus, which can confuse the receiver who is asked to digest information and consider advice that can sometimes be contradictory and even inaccurate. It is therefore important that details regarding specific information sources are clearly provided and properly referenced so that the grower is in a position to access and assess the quality of the information and act on it accordingly. A neat solution to these issues would be a coordinated and coherent set of simple IPM messages presented through established and respected communication bodies. Whereas this would be an ideal solution it is not likely to become a reality because different stakeholders communicate different messages to suit their specific needs, whether that be the selling of pesticides or seed, a reduction in pesticide contamination of waterways or an increase in local biodiversity. This point is illustrated by the fact that there are now multiple different definitions of IPM (Prokopsy, 2003), each abiding by the same principle of a holistic approach to pest management that attempts to minimise the negative impacts of pest management programmes on the environment (Parsa et al., 2014). However, slight variations in definition are employed to suit the end goal of the authors. Farmers too may also disagree as to what constitutes best IPM practices as they have different drivers to uptake of IPM practices which may be a combination of economic, environmental and social (Sherman and Gent, 2014). Despite such differences in opinion, the main components of IPM programmes and the potential benefits of adopting them for the grower are often fairly well conserved among practitioners, which is important as crop

© Burleigh Dodds Science Publishing Limited, 2020. All rights reserved.

protection decisions often involve multiple decision makers (Burnett and Hughes, 2015). The question is whether adequate evidence-based advice supporting IPM adoption currently exists, and if it does, whether the messages encouraging adoption are being communicated effectively.

Recent research has shown that growers who perceive themselves to be more familiar with IPM are more actively seeking IPM information citing knowledge exchange activities, via open days, crop walks, discussion groups, exchanges with independent agronomists, as being their most valued sources of pest management advice (Creissen et al., 2018). Those growers who attend knowledge exchange events are, in turn, more likely to adopt IPM practices on farm (Creissen et al., 2018). The concern is that those individuals who are actively engaging with experts and peers to seek and understand current evidence-based advice are not in the majority. That presents an interesting question, how can we incentivise growers to engage in knowledge exchange activities? There are various 'carrots' and 'sticks' that can be used to this end. A 'carrot' may be in the form of financial rewards for attending knowledge exchange events. In Ireland growers are rewarded to the sum of €750 per annum for their attendance at five discussion group meetings per annum and completion of associated tasks such as a farm improvement plan (https://www.teagasc.ie/about/farm-advisory/advisory-services/discussion-groups/). 'Sticks' to incentivise uptake of IPM include subsidy reductions and fines for those not deemed to be practising IPM. This is a less desirable route as it requires the development of system to assess whether IPM is being adopted and IPM approaches will vary greatly according to the farm business, farming system and the associated biotic threats to crop production. Many potential 'stick or 'carrot' systems would require an effective way of measuring the complex technology that is IPM to be established. Recent efforts to establish a universal metric for assessing IPM uptake have been made (Creissen et al. 2019). If accepted by the industry this could pave the way for a system that could be employed to encourage farmers to increase their level of IPM adoption.

When considering malting barley production we must also be aware of the high level of influence of those involved in the supply chain (Barnes et al., 2018). Distillers aim to acquire malt that provides the greatest alcohol spirit yield, brewers may also consider the flavour profile of the malt. Maltsters specify grain quality parameters which often relate to uniformity of germination rates and biochemistry of the grains in relation to malting ability, such traits include specific grain weight and degree of grain skinning (Hoad et al., 2016). To meet these requirements growers look to maximise grain fill and minimise skinning which are both affected by fungicides and growth regulators (Wilson et al., 2018). For a high-value crop this may reduce the likelihood of growers accepting potential risks associated with a reduction in inputs. Due to the need to grow for the market, farmers are also heavily restricted in the cultivars they may grow as the end users have already pre-selected a few cultivars they would be willing to accept. Removal of choices associated with this key and fundamental component of any disease management programme means that growers must alleviate risks associated with disease by other means, including the use of fungicides. This presents a considerable barrier to uptake of IPM in the malting barley sector. However, for winter barley the issues surrounding heterogeneity of the end produce are far less significant as nearly all of the grain produced is intended for livestock feed. Research into the potential for, and barriers to, IPM uptake along the entire supply chain is critical to improving our understanding of the most effective approaches to communicate the IPM message.

© Burleigh Dodds Science Publishing Limited, 2020. All rights reserved.

10 Farming systems, soil and research platforms

As an integrated approach, both the crop and soil environment are the product of multiple components spread through time. Factorial experimentation can be used only to evaluate the possible contribution of a few components and their interactions. IPM must be assessed as a system in comparison with other systems. Therefore, it is not possible to do this for barley alone as an IPM system will have barley as a crop in a sequence. Nevertheless, IPM schemes have been compared in sequences not only across multiple sites but also across years in long-term experiment platforms which reveals where they work well and where their weaknesses are (Lechenet et al., 2016; Vasileiadis et al., 2017).

Such research platforms are helpful and while few of these focus on barley, the principles still apply. However, in common with cultivar testing and recommendation trials, they are seldom representative of on-farm conditions. On the other hand, on-farm conditions are highly variable so it is difficult to conceive what representative trial characteristics should comprise. Fundamentally the requirement for robust IPM on-farm is resilience, so trialling systems that express biotic and abiotic stress factors as a proxy for on-farm conditions to identify reliable IPM protocols are what is needed. Borrowing concepts used particularly in disciplines like molecular biology, an indication of the key components of a system, can be gained by overexpression or knock-out of that factor (Newton, 2016). Much criticised by systems biologists, nevertheless such approaches can indicate the sensitive or rate-limiting factors in IPM systems that should be the focus of research systems comparisons. For example, zero tillage and substantially reduced nutrients could be used as an overexpression/knock-out proxy for on-farm conditions (Newton et al., 2018) to identify cultivars likely to be resilient. Tolerance, essentially the expression of yield resilience with respect to disease, can be characterised in this way (Newton, 2016).

We have discussed briefly plant–microbe management in the broad as well as IPM-applied context including endophytes and mycorrhizal colonisation. We must also reference the microbiome in general and perhaps the rhizosphere biome in particular. The ability of many soils to be disease suppressive is well known, as is the ability of bacteria in particular to induce resistance or induced systemic resistance (ISR) (Bakker et al., 2013) and plant growth-promoting rhizobacteria. With modern sequencing technologies we are beginning to obtain data on the importance of microbiome characteristics, but we have yet to understand sufficiently their function and how we might manage and manipulate them for beneficial outcomes. Crucial to any microbial activity in the soil, however, is soil management and cultivation in particular as soil disturbance can, for example, disrupt mycorrhizal networks while enhancing bacterial oxidative processes.

Implied in the above discussion of uptake of IPM is the incremental adoption of measures that should improve crop health. A high risk but potentially much more rewarding strategy is the adoption of multiple measures together in a systems approach. Although there is a transition phase to successful implementation, the outcome cannot be achieved using an incremental approach. An example is intercropping in a conservation tillage system. Here several technologies must be adopted at once. Inversion soil tillage is replaced by non-inversion tillage (minimum or zero tillage/direct drilling). Soil structure and fertility is managed by using cover crops. Disease is managed by a combination of resistance in the crop cultivars, diversity and deployment strategies. Nutrient management is further managed in this context taking due regard for each component of the crop system. The

© Burleigh Dodds Science Publishing Limited, 2020. All rights reserved.

barley crop itself is grown together with another species, often a legume such as peas. Harvest dates must be matched and subsequent yield separation may be needed if used in different markets such as for malting. This can be achieved and examples of an approach as the one described have been achieved on farm even in challenging winter environments in Scotland and demonstrated to other farmers to encourage adoption. Therefore, to properly implement IPM in barley the whole system should be considered before decisions are made about adoption to determine the best approach or direction of travel for any given situation.

11 Conclusion and future trends

For increased IPM adoption in barley production we need continued research and innovation, but equally important is translation of the research outcomes into practical implementation and communication of these outcomes to farmers. However, to both ensure the practicality and that the solutions are needed, all IPM research and development should be carried out in a co-production/co-innovation framework. Indeed, many practical implementations of IPM components are being developed by farmers both deliberately and inadvertently through a practical understanding of what works in practice. This needs to inform more formal experimentation and innovation. Discussion groups, field labs, monitor farms, farm and trial visits, open days and other groupings of farmers together with researchers and advisors can all help achieve these outcomes, but coordinated and collaborative trialling and demonstration, especially if facilitated by funding, can be particularly effective.

   IPM involves multiple decision makers that includes farmers, agronomists and other farm business experts, and all have different attitudes towards risk driven by their personal nature, circumstances and their business environment. However, IPM has a major public goods driver too in the form of the environment and long-term sustainability. This needs to be separately incentivised, as does the development of all the tools needed to implement effective IPM to deliver these goods.

12 Acknowledgements

We are grateful for the financial support for this work from the Rural and Environmental Science and Analytical Services Division of the Scottish Government under its Environmental Change and Food, Land and People Research Programmes.

13 Where to look for further information

13.1 Further reading

- A good review of control of foliar diseases of barley is Walters et al. (2012).
- The broader context of barley is reviewed by Newton et al. (2011).

© Burleigh Dodds Science Publishing Limited, 2020. All rights reserved.

13.2 Key conferences

- International Congress for Plant Pathology, Lyon, France, 2023.
- Many countries host national agronomy and pathology meetings on IPM, for example check the Association of Applied Biologists website.

13.3 Major international research projects

- The EU hosted several large IPM projects that have legacy websites, ENDURE http://www.endure-network.eu/, and PURE, http://www.pure-ipm.eu/being the major ones.

14 References

AHDB. 2016. Agriculture and Horticulture Development Board barley disease management guide. Available at: https://cereals.ahdb.org.uk/publications/2016/february/23/barley-disease-management-guide.aspx (accessed November 2018).

AHDB. 2018. Agriculture and Horticulture Development Board recommended lists. Available at: https://cereals.ahdb.org.uk/varieties/ahdb-recommended-lists.aspx (accessed November 2018).

Alizadeh, A., Benetti, V., Sarrafi, A., Barrault, G. and Albertini, L. 1994. Genetic-analysis for partial resistance to an Iranian strain of bacterial leaf streak (*Xanthomonas campestris* pv *hordei*) in barley. *Plant Breed.* 113, 323–6.

Bai, G. H. and Shaner, G. 2004. Management and resistance in wheat and barley to Fusarium head blight. *Annu. Rev. Phytopathol.* 42, 135–61. doi:10.1146/annurev.phyto.42.040803.140340.

Bakker, P. A. H. M., Doornbos, R. F., Zamioudis, C., Berendsen, R. L. and Pieterse, C. M. J. 2013. Induced systemic resistance and the rhizosphere microbiome. *Plant Pathol. J.* 29(2), 136–43. doi:10.5423/PPJ.SI.07.2012.0111.

Balba, H. 2007. Review of strobilurin fungicide chemicals. *J. Environ. Sci. Health B* 42(4), 441–51. doi:10.1080/03601230701316465.

Barnes, A., Burnett, F. and Creissen, H. 2018. Assessing the barriers to uptake of Integrated Pest Management in Scotland. Available at: https://www.sruc.ac.uk/downloads/file/3764/assessing_the_barriers_to_uptake_of_integrated_pest_management_in_scotland (accessed November 2018).

Bateman, G. L., Gutteridge, R. J., Gherbawy, Y., Thomsett, M. A. and Nicholson, P. 2007. Infection of stem bases and grains of winter wheat by *Fusarium culmorum* and *F. graminearum* and effects of tillage method and maize-stalk residues. *Plant Pathol.* 56(4), 604–15. doi:10.1111/j.1365-3059.2007.01577.x.

Bingham, I. J. and Newton, A. C. 2009. Crop tolerance of foliar pathogens: possible mechanisms and potential for exploitation. In: Walters, D. (Ed.), *Disease Control in Crops–Biological and Environmentally Friendly Approaches*. Wiley-Blackwell, Chichester, UK, pp. 142–61.

Bingham, I. J. and Topp, C. F. E. 2009. Potential contribution of selected canopy traits to the tolerance of foliar disease by spring barley. *Plant Pathol.* 58(6), 1010–20. doi:10.1111/j.1365-3059.2009.02137.x.

Bingham, I. J., Walters, D. R., Foulkes, M. J. and Paveley, N. D. 2009. Crop traits and the tolerance of wheat and barley to foliar disease. *Ann. Appl. Biol.* 154(2), 159–73. doi:10.1111/j.1744-7348.2008.00291.x.

Bockus, W. W. and Claassen, M. M. 1992. Effects of crop-rotation and residue management-practices on severity of tan spot of winter-wheat. *Plant Dis.* 76(6), 633–6. doi:10.1094/PD-76-0633.

Braun-Kiewnick, A., Jacobsen, B. J. and Sands, D. C. 2000. Biological control of *Pseudomonas syringae* pv. *syringae*, the causal agent of basal kernel blight of barley, by antagonistic *Pantoea agglomerans*. *Phytopathology* 90(4), 368–75. doi:10.1094/PHYTO.2000.90.4.368.

© Burleigh Dodds Science Publishing Limited, 2020. All rights reserved.

Brooker, R. W., Karley, A. J., Newton, A. C., Pakeman, R. J. and Schöb, C. 2016. Facilitation and sustainable agriculture: a mechanistic approach to reconciling crop production and conservation. *Funct. Ecol.* 30(1), 98–107. doi:10.1111/1365-2435.12496.

Browne, R. A. and Cooke, B. M. 2005. A comparative assessment of potential components of partial disease resistance to Fusarium head blight using a detached leaf assay of wheat, barley and oats. *Eur. J. Plant Pathol.* 112(3), 247–58. doi:10.1007/s10658-005-2077-z.

Burnett, F. and Hughes, G. 2004. The development of a risk assessment method to identify wheat crops at risk from eyespot. AHDB project report 347. AHDB, Stoneleigh Park, Kenilworth, Warwickshire, UK.

Burnett, F., Butler-Ellis, C., Hughes, G., Knight, S. and Rumiana, R. 2012. Forecasting eyespot development and yield losses in winter wheat. AHDB project report 491. AHDB, Stoneleigh Park, Kenilworth, Warwickshire, UK.

Byerlee, D. 1996. Modern varieties, productivity, and sustainability—recent experience and emerging challenges. *World Dev.* 24(4), 697–718. doi:10.1016/0305-750X(95)00162-6.

Carignano, M., Staggenborg, S. A. and Shroyer, J. P. 2008. Management practices to minimize tan spot in a continuous wheat rotation. *Agron. J.* 100, 145–53.

Caten, C. E. and Newton, A. C. 2000. Variation in cultural characteristics, pathogenicity, vegetative compatibility and electrophoretic karyotype within field populations of *Stagonospora nodorum*. *Plant Pathol.* 49(2), 219–26. doi:10.1046/j.1365-3059.2000.00441.x.

Chandler, D., Bailey, A. S., Tatchell, G. M., Davidson, G., Greaves, J. and Grant, W. P. 2011. The development, regulation and use of biopesticides for integrated pest management. *Philos. Trans. R. Soc. Lond., B, Biol. Sci.* 366(1573), 1987–98. doi:10.1098/rstb.2010.0390.

Chen, F. Q., Prehn, D., Hayes, P. M., Mulrooney, D., Corey, A. and Vivar, H. 1994. Mapping genes for resistance to barley stripe rust (*Puccinia striiformis* f. sp. *hordei*). *Theor. Appl. Genet.* 88(2), 215–9. doi:10.1007/BF00225900.

Colbach, N., Lucas, P., Cavelier, N. and Cavelier, A. 1997. Influence of cropping system on sharp eyespot in winter wheat. *Crop Prot.* 16(5), 415–22. doi:10.1016/S0261-2194(97)00018-5.

Cooke, L. R., Locke, T., Lockley, K. D., Phillips, A. N., Sadiq, M. D. S., Coll, R., Black, L., Taggart, P. J. and Mercer, P. C. 2004. The effect of fungicide programmes based on epoxiconazole on the control and DMI sensitivity of *Rhynchosporium secalis* in winter barley. *Crop Prot.* 23(5), 393–406. doi:10.1016/j.cropro.2003.09.009.

Creissen, H., Tranter, R., Jones, P., Burnett, F., Jess, J., Girling, R., Gaffney, M., Thorne, F. and Kildea, S. 2018. IPM practices on arable farms in the UK and Ireland. In: *Crop Production in Northern Britain*, Dundee Conference, UK, 2018.

Creissen, H. E., Jones, P. J., Tranter, R. B., Girling, R. D., Jess, S., Burnett, F. J., Gaffney, M., Thorne, F. S. and Kildea, S. 2019. Measuring the unmeasurable? A method to quantify adoption of Integrated Pest Management practices in temperate arable farming systems. *Pest. Man. Sci.* doi: 10.1002/ps.5428

Cromey, M. G., Parkes, R. A. and Fraser, P. M. 2006. Factors associated with stem base and root diseases of New Zealand wheat and barley crops. *Aust. Plant Pathol.* 35, 391–400.

Duczek, L. J., Sutherland, K. A., Reed, S. L., Bailey, K. L. and Lanford, G. P. 1999. Survival of leaf spot pathogens on crop residues of wheat and barley in Saskatchewan. *Can. J. Plant Pathol.* 21, 165–73.

Edwards, S. 2013. HGCA risk assessment for Fusarium mycotoxins in wheat. Caledonia House: Home Grown Cereals Authority Topic Sheet 121. AHDB, Stoneleigh Park, Kenilworth, Warwickshire, UK.

El Attari, H., Hayes, P. M., Rebai, A., Barrault, G., Dechamp-Guillaume, G. and Sarrafi, A. 1998. Potential of doubled-haploid lines and localization of quantitative trait loci (QTL) for partial resistance to bacterial leaf streak (*Xanthomonas campestris* pv. *hordei*) in barley. *Theor. Appl. Gen.* 96(1), 95–100. doi:10.1007/s001220050714.

European Union. 2009. Council Directive 2009/128/EC of the European Parliament and of the Council of 21 October 2009 establishing a framework for Community action to achieve the sustainable use of pesticides. *Off. J. Eur. Comm.* L309/71, 71–86.

© Burleigh Dodds Science Publishing Limited, 2020. All rights reserved.

Fabre, F., Pierre, J. S., Dedryver, C. A. and Plantegenest, M. 2006. Barley yellow dwarf virus disease risk assessment based on Bayesian modelling of aphid population dynamics. *Ecol. Modell.* 193(3–4), 457–66. doi:10.1016/j.ecolmodel.2005.08.021.

FAO. 2018. Food and Agriculture Organization of the United Nations (FAO) definition of Integrated Pest Management (IPM). Available at: http://www.fao.org/agriculture/crops/thematic-sitemap/theme/pests/ipm/en/(accessed on 4 August 2018).

FERA. 2018. Crop monitor. Available at: http://www.cropmonitor.co.uk/ (accessed on 16 September 2018).

FRAC Code List. 2018. Fungicides sorted by mode of action (including FRAC Code numbering). Available at: http://www.frac.info/docs/default-source/publications/frac-code-list/frac_code_list_2018-final.pdf?sfvrsn=6144b9a_2.

Geiger, F., Bengtsson, J., Berendse, F., Weisser, W. W., Emmerson, M., Morales, M. B., Ceryngier, P., Liira, J., Tscharntke, T., Winqvist, C., et al. 2010. Persistent negative effects of pesticides on biodiversity and biological control potential on European farmland. *Basic Appl. Ecol.* 11(2), 97–105. doi:10.1016/j.baae.2009.12.001.

Grasso, V., Sierotzki, H., Garibaldi, A. and Gisi, U. 2006. Characterization of the cytochrome *b* gene fragment of *Puccinia* species responsible for the binding site of QoI fungicides. *Pest. Biochem. Physiol.* 84(2), 72–82. doi:10.1016/j.pestbp.2005.05.005.

Grimmer, M. K., van den Bosch, F., Powers, S. J. and Paveley, N. D. 2015. Fungicide resistance risk assessment based on traits associated with the rate of pathogen evolution. *Pest Manag. Sci.* 71(2), 207–15. doi:10.1002/ps.3781.

Havis, N. D. and Gorniak, K. 2014. Cephalosporium leaf stripe – an emerging threat to wheat crops in short rotations. AHDB project report 542. AHDB, Stoneleigh Park, Kenilworth, Warwickshire, UK.

Havis, N. D., Burnett, F. J., Hughes, G., Mercer, P. C., Cooke, L. R., Fraaije, B. A., Hunter, E. A. and Oxley, S. J. P. 2014a. *Rhynchosporium commune* – understanding the effects of variety, fungicide resistance and seed-borne infection on disease levels in barley. In: *The Dundee Conference. Crop Protection in Northern Britain 2014*, Dundee, UK, pp. 167–72.

Havis, N. D., Nyman, M. and Oxley, S. J. P. 2014b. Evidence for seed transmission and symptomless growth of *Ramularia collo-cygni* in barley (*Hordeum vulgare*). *Plant Pathol.* 63(4), 929–36. doi:10.1111/ppa.12162.

Havis, N. D., Brown, J. K. M., Clemente, G., Frei, P., Jedryczka, M., Kaczmarek, J., Kaczmarek, M., Matusinsky, P., McGrann, G. R. D., Pereyra, S., et al. 2015. *Ramularia collo-cygni* – an emerging pathogen of barley crops. *Phytopathology* 105(7), 895–904. doi:10.1094/PHYTO-11-14-0337-FI.

Havis, N. D., Gorniak, K., Taylor, J., Stanisz-Migal, M. and Burnett, F. J. 2018. Controlling Ramularia leaf spot in barley. In: *Therapeutic Dundee Conference. Crop Production in Northern Britain 2018*, Dundee, UK, pp. 91–6.

Hoad, S. P., Brennan, M., Wilson, G. W. and Cochrane, P. M. 2016. Hull to caryopsis adhesion and grain skinning in malting barley: identification of key growth stages in the adhesion process. *J. Cer. Sci.* 68, 8–15. doi:10.1016/j.jcs.2015.10.007.

Holland, J. E., Bennett, A., Newton, A. C., White, P., McKenzie, B., George, T., Pakeman, R., Bailey, J., Fornara, D. and Hayes, R. 2017. Liming impacts on soils, plants and biodiversity in the UK: a review. *Sci. Total Environ.* 610–611, 316–32.

Hughes, G., McRoberts, N. and Burnett, F. J. 1999. Decision making and diagnosis in disease management. *Plant Pathol.* 48(2), 147–53. doi:10.1046/j.1365-3059.1999.00327.x.

Hughes, G. and Burnett, F., J. 2015. Integrating experience, evidence and expertise in the crop protection decision process. *Plant Disease*, Vol 99 No 9, 1197–1203.

Jørgensen, L. N., van den Bosch, F., Oliver, R. P., Heick, T. M. and Paveley, N. D. 2017. Targeting fungicide inputs according to need. *Annu. Rev. Phytopathol.* 55, 181–203. doi:10.1146/annurev-phyto-080516-035357.

Kendall, S., Hollomon, D. W., Ishii, H. and Heaney, S. P. 1994. Characterization of benzimidazole-resistant strains of *Rhynchosporium secalis*. *Pestic. Sci.* 40(3), 175–81. doi:10.1002/ps.2780400302.

© Burleigh Dodds Science Publishing Limited, 2020. All rights reserved.

Kennedy, S. P., Bingham, I. J. and Spink, J. H. 2017. Determinants of spring barley yield in a high yield potential environment. *J. Agric. Sci.* 155(1), 60–80. doi:10.1017/S0021859616000289.

Kergunteuil, A., Bakhtiari, M., Formenti, L., Xiao, Z., Defossez, E. and Rasmann, S. 2016. Biological control beneath the feet: a review of crop protection against insect root herbivores. *Insects* 29(4). pii:E70. doi:10.3390/insects7040070.

King, K. M., West, J. S., Brunner, P. C., Dyer, P. S. and Fitt, B. D. L. 2013. Evolutionary relationships between *Rhynchosporium lolii* sp. nov. and other *Rhynchosporium* species on grasses. *PLOS ONE* 8(10), e72536. doi:10.1371/journal.pone.0072536.

Kleinhofs, A., Brueggeman, R., Nirmala, J., Zhang, L., Mirlohi, A., Druka, A., Rostoks, N. and Steffenson, B. J. 2009. Barley stem rust resistance genes: structure and function. *Plant Genome* 2(2), 109–20. doi:10.3835/plantgenome2009.02.0011.

Kusch, S. and Panstruga, R. 2017. *mlo*-based resistance: an apparently universal 'weapon' to defeat powdery mildew disease. *Mol. Plant Microbe Int.* 30, 179–89.

Lamichhane, J. R., Dachbrodt-Saaydeh, S., Kudsk, P. and Messean, A. 2016. Toward a reduced reliance on conventional pesticides in European agriculture. *Plant Dis.* 100(1), 10–24. doi:10.1094/PDIS-05-15-0574-FE.

Lechenet, M., Deytieux, V., Antichic, D., Aubertot, J.-N., Bàrberi, B., Bertrand, M., Cellier, V., Charles, R., Colnenne-David, C., Dachbrodt-Saaydeh, S., et al. 2016. Diversity of methodologies to experiment Integrated Pest Management in arable cropping systems: analysis and reflections based on a European network. *Eur. J. Agron.* 83, 86–99.

Looseley, M. E. and Newton, A. C. 2014. Assessing the consequences of microbial infection in field trials: seen, and unseen, beneficial, parasitic and pathogenic. *Agronomy* 4(2), 302–21. doi:10.3390/agronomy4020302.

Looseley, M. E., Newton, A. C., Atkins, S. D., Fitt, B. D. L., Fraaije, B. A., Thomas, W. T. B., Keith, R., Macaulay, M., Lynott, J. and Harrap, D. 2012. Genetic basis of control of *Rhynchosporium secalis* infection and symptom expression in barley. *Euphytica* 184(1), 47–56. doi:10.1007/s10681-011-0485-z.

Lyngkjaer, M. F., Newton, A. C., Atzema, J. L. and Baker, S. J. 2000. The barley *mlo*-gene: an important powdery mildew resistance source. *Agronomie* 20(7), 745–56. doi:10.1051/agro:2000173.

Mathre, D. E. 1997. *Compendium of Barley Diseases* (2nd edn.). The American Phytopathological Society Press, St Paul, MN.

Matthews, G. A. 2000. *Pesticide Application Methods* (3rd edn.). Blackwell Publishing Science, Oxford, UK.

McArt, S. H., Urbanowicz, C., McCoshum, S., Irwin, R. E. and Adler, L. S. 2017. Landscape predictors of pathogen prevalence and range contractions in US bumblebees. *Proc. Biol. Sci.* 284(1867). doi:10.1098/rspb.2017.2181:29142119.

Menzies, J. G., Turkington, T. K. and Knox, R. E. 2009. Testing for resistance to smut diseases of barley, oats and wheat in western Canada. *Can. J. Plant Pathol.* 31(3), 265–79. doi:10.1080/07060660909507601.

Menzies, J. G., Steffenson, B. J. and Kleinhofs, A. 2010. A resistance gene to *Ustilago nuda* in barley is located on chromosome 3H. *Can. J. Plant Pathol.* 32(2), 247–51. doi:10.1080/07060661003739977.

Miller, W. A. and Rasochova, L. 1997. Barley yellow dwarf viruses. *Ann. Rev. Phytopathol.* 35, 167–90.

Mississippi State University. 2018. Mississippi crop situation. Available at: http://www.mississippi-crops.com/disease-monitoring/.

Mundt, C. C. 2002. Use of multiline cultivars and cultivar mixtures for disease management. *Annu. Rev. Phytopathol.* 40, 381–410. doi:10.1146/annurev.phyto.40.011402.113723.

Newton, A. C. 1989. Genetic adaptation of *Erysiphe graminis* f.sp. *hordei* to barley with partial resistance. *J. Phytopathol.* 126(2), 133–48. doi:10.1111/j.1439-0434.1989.tb01097.x.

Newton, A. C. 2016. Exploitation of diversity within crops – the key to disease tolerance? *Front. Plant Sci.* 7, 665. doi:10.3389/fpls.2016.00665.

Newton, A. C. and Guy, D. C. 2009. The effects of uneven, patchy cultivar mixtures on disease control and yield in winter barley. *Field Crops Res.* 110(3), 225–8. doi:10.1016/j.fcr.2008.09.002.

© Burleigh Dodds Science Publishing Limited, 2020. All rights reserved.

Newton, A. C. and Guy, D. C. 2011. Scale and spatial structure effects on the outcome of barley cultivar mixture trials for disease control. *Field Crops Res.* 123(2), 74–9. doi:10.1016/j.fcr.2011.05.002.

Newton, A. C., Ellis, R. P., Hackett, C. A. and Guy, D. C. 1997. The effect of component number on *Rhynchosporium secalis* infection and yield in mixtures of winter barley cultivars. *Plant Pathol.* 46(6), 930–8. doi:10.1046/j.1365-3059.1997.d01-83.x.

Newton, A. C., Begg, G. S. and Swanston, J. S. 2009. Deployment of diversity for enhanced crop function. *Ann. Appl. Biol.* 154(3), 309–22. doi:10.1111/j.1744-7348.2008.00303.x.

Newton, A. C., Fitt, B. D. L., Atkins, S. D., Walters, D. R. and Daniell, T. 2010. Pathogenesis, mutualism and parasitism in the trophic space of microbe-plant interactions. *Trends Microbiol.* 18, 365–73.

Newton, A. C., Flavell, A. J., George, T. S., Leat, P., Mullholland, B., Ramsay, L., Revoredo-Giha, C., Russell, J., Steffenson, B. J., Swanston, J. S., et al. 2011. Crops that feed the world 4. Barley: a resilient crop? Strengths and weaknesses in the context of food security. *Food Sec.* 3(2), 141–78. doi:10.1007/s12571-011-0126-3.

Newton, A. C., Guy, D. C., Valentine, T., George, T. S., McKenzie, B. M. and Hackett, C. A. 2018. Differential adaptation of spring barley cultivars to inversion and non-inversion tillage. In: *Proceedings, Crop Production in Northern Britain 2018: The Dundee Conference, Environmental Management and Crop Production*, pp. 109–14.

O'Grady, M. J. and O'Hare, G. M. P. 2017. Modelling the smart farm Information. *Proc. Agric.* 4, 179–87.

Oliver, R. P. and Hewitt, H. G. 2014. *Fungicides in Crop Protection* (2nd edn.). CABI Publishing, CAB International, Wallingford, UK.

Orlando, B., Maumene, C. and Piraux, F. 2017. Ergot and ergot alkaloids in French cereals: occurrence, pattern and agronomic practices for managing the risk. *World Mycotoxin J.* 10(4), 327–38. doi:10.3920/WMJ2017.2183.

Oxley, S. J. P. and Burnett, F. 2009. *Barley Disease Control.* SAC Technical Note TN619. SAC, Edinburgh, UK. ISBN: 1 85482 873 8.

Parsa, S., Morse, S., Bonifacio, A., Chancellor, T. C. B., Condori, B., Crespo-Perez, V., Hobbs, S. L. A., Kroschel, J., Ba, M. N., Rebaudo, F., et al. 2014. Obstacles to integrated pest management adoption in developing countries. *Proc. Natl. Acad. Sci. U.S.A.* 111(10), 3889–94. doi:10.1073/pnas.1312693111.

Pecchioni, N., Vale, G., Toubia-Rahme, H., Faccioli, P., Terzi, V. and Delogu, G. 1999. Barley-*Pyrenophora graminea* interaction: QTL analysis and gene mapping. *Plant Breed.* 118(1), 29–35. doi:10.1046/j.1439-0523.1999.118001029.x.

Penselin, D., Münsterkötter, M., Kirsten, S., Felder, M., Taudien, S., Platzer, M., Ashelford, K., Paskiewicz, K. H., Harrison, R. J., Hughes, D. J., et al. 2016. Comparative genomics to explore phylogenetic relationship, cryptic sexual potential and host specificity of *Rhynchosporium* species on grasses. *BMC Genomics* 17(1), 953, doi:10.1186/s12864-016-3299-5.

Prokopy, R. J. 2003. Two decades of bottom-up, ecologically based pest management in a small commercial apple orchard in Massachusetts. *Agric. Ecosyst. Environ.* 94(3), 299–309. doi:10.1016/S0167-8809(02)00036-1.

Scherm, B., Balmas, V., Spanu, F., Pani, G., Delogu, G., Pasquali, M. and Migheli, Q. 2013. *Fusarium culmorum*: causal agent of foot and root rot and head blight on wheat. *Mol. Plant Pathol.* 14(4), 323–41. doi:10.1111/mpp.12011.

Scottish Government. 2017. Pesticide usage in Scotland: Arable crops and potato stores 2016. Appendix 6 – Integrated pest management. Available at: https://www.gov.scot/Publications/2 017/10/7179/59 (accessed November 2018).

Semar, M., Strobel, D., Koch, A., Klappach, K. and Stammler, G. 2007. Field efficacy of pyraclostrobin against populations of *Pyrenophora teres* containing the F129L mutation in the cytochrome *b* gene. *J. Plant Dis. Prot.* 114(3), 117–9. doi:10.1007/BF03356718.

Sherman, J. and Gent, D. H. 2014. Concepts of sustainability, motivations for pest management approaches, and implications for communicating change. *Plant Dis.* 98(8), 1024–35. doi:10.1094/PDIS-03-14-0313-FE.

© Burleigh Dodds Science Publishing Limited, 2020. All rights reserved.

UKCPVS. 2018. The United Kingdom cereal pathogen virulence survey. Available at: https://cereals. ahdb.org.uk/ukcpvs.

van den Bosch, F., Oliver, R., van den Berg, F. and Paveley, N. 2014a. Governing principles can guide fungicide resistance management tactics. *Annu. Rev. Phytopathol.* 52, 175–95. doi:10.1146/ annurev-phyto-102313-050158.

van den Bosch, F., Paveley, N., van den Berg, F., Hobbelen, P. and Oliver, R. 2014b. Mixtures as a fungicide resistance management tactic. *Phytopathology* 104(12), 1264–73. doi:10.1094/ PHYTO-04-14-0121-RVW.

Vasileiadis, V. P., Dachbrodt-Saaydeh, S., Kudsk, P., Colnenne-David, C., Leprince, F., Holb, I. J., Kierzek, R., Furlan, L., Loddo, D., Melander, B., et al. 2017. Sustainability of European winter wheat- and maize-based cropping systems: economic, environmental and social *ex-post* assessment of current and IPM-based systems. *Crop Prot.* 97, 60–9.

Walters, D. R. and Bingham, I. J. 2007. Influence of nutrition on disease development caused by fungal pathogens: implications for plant disease control. *Ann. Appl. Biol.* 151(3), 307–24. doi:10.1111/j.1744-7348.2007.00176.x.

Walters, D. R., Havis, N. D. and Oxley, S. J. P. 2008. *Ramularia collo-cygni*: the biology of an emerging pathogen of barley. *FEMS Microbiol. Lett.* 279(1), 1–7. doi:10.1111/j.1574-6968.2007.00986.x.

Walters, D. R., Avrova, A., Bingham, I. J., Burnett, F. J., Fountaine, J., Havis, N. D., Hoad, S. P., Hughes, G., Looseley, M., Oxley, S. J. P., et al. 2012. Control of foliar diseases in barley: towards an integrated approach. *Eur. J. Plant Pathol.* 133(1), 33–73. doi:10.1007/s10658-012-9948-x.

Walters, D. R., Lyon, G. D. and Newton, A. C. 2014. *Induced Resistance for Plant Defence: a Sustainable Approach to Crop Protection* (2nd edn.). Blackwell Publishing Science, Oxford, UK.

Wilson, W., Dahl, B. and Nganje, W. 2018. Economic costs of Fusarium Head Blight, scab and deoxynivalenol. *World Mycotoxin J.* 11(2), 291–302. doi:10.3920/WMJ2017.2204.

Xu, X. M., Jeffries, P., Pautasso, M. and Jeger, M. J. 2011. Combined use of biocontrol agents to manage plant diseases in theory and practice. *Phytopathology* 101(9), 1024–31. doi:10.1094/ PHYTO-08-10-0216.

Young, C. S., Thomas, J. M., Parker, S. R. and Paveley, N. D. 2006. Relationship between leaf emergence and optimum spray timing for leaf blotch (*Rhynchosporium secalis*) control on winter barley. *Plant Pathol.* 55(3), 413–20. doi:10.1111/j.1365-3059.2006.01361.x.

Yan, G. P. and Chen, X. M. 2006. Molecular mapping of a recessive gene for resistance to stripe rust in barley. *Theor. Appl. Genet.* 113(3), 529–37. doi:10.1007/s00122-006-0319-x.

Zhan, J., Fitt, B. D. L., Pinnschmidt, H. O., Oxley, S. J. P. and Newton, A. C. 2008. Resistance, epidemiology and sustainable management of *Rhynchosporium secalis* populations on barley. *Plant Pathol.* 57, 1–14.

© Burleigh Dodds Science Publishing Limited, 2020. All rights reserved.

Chapter 4

Diseases affecting grain legumes and their management

Keith Thomas, University of Sunderland, UK

1 Introduction

All field crops suffer from diseases of some sort, often in combination. This is increasingly so in monoculture growth where pathogen evolution may be out of synchrony with crop development and new crop varieties push mutualistic microorganisms towards parasitism.

Moreover, as modern agriculture is increasingly pressured to improve productivity, breeding is incentivised to select for traits associated with yield and character. Traits to resist disease may be judged of lower importance if other means of controlling disease are available. For many years, the proliferation and effectiveness of chemical control allowed this, but with recent moves to limit exposure to chemicals, the priorities of other means of disease resistance have increased. This is, of course, a difficult balance. Pressures for yield are motivated by the justifiable need to satisfy food demand, but in recent years awareness of pesticide hazards have tempered their use. Calculations that pesticide poisoning can cause more deaths than infectious diseases in the developing world (Eddleston et al., 2002) have indicated that uncontrolled means of achieving food security cannot justify the target.

Increases in disease may be seen as a consequence of changes in pathogen or host physiology or through a trade-off between physiological systems where resistance against one disease reduces resistance to others (McGrann et al., 2014; Brown and Rant 2013). Progressive legume cultivation is no different but has specific features relevant to disease development as well as a range of adapted pathogens with both biotrophic and necrotrophic consequences.

http://dx.doi.org/10.19103/AS.2017.0023.10
© Burleigh Dodds Science Publishing Limited, 2018. All rights reserved.

Climate change contributes additional variabilities to the anticipation of disease and is a further element in food insecurity (Ghini et al., 2008; Chakroborty et al., 2000) and financial cost. Losses to soya bean crops by rust are estimated to have cost up to $2 billion between 1996 and 2007 (Wrather and Koenning, 2009). Losses in other continents are similarly severe (Murithi et al., 2015), with up to 15% of food legume production in India lost to disease (Pandey et al., 2009).

In some cases, disease is a local concern often in small-scale production, possibly due to poor management of soils, infected seeds or lack of crop rotation. In other cases, large-scale production may be threatened by regional or epidemic disease spread as when a virulent strain of a pathogen arises and spreads without warning. Obviously, these two overlap and may cross infect, thus requiring a broad approach as advocated by integrated management systems (Davidson and Kimber, 2007).

2 Grain legume diseases

Grain legumes incur infection by broad pathogen groups such as *Fusarium* and *Botrytis* as well as more species-specific infections from *Colletrotrichum* and *Ascochyta* fungi. Bacterial blight from *Pseudomonas syringae* and a range of viral infections are also found in many crop areas. Table 1 lists the range of these diseases, but it is by no means exhaustive. The incidence and severity of each disease depends, though, on local climate and circumstance and on the host legume. Nevertheless, some diseases stand out for their prevalence. For example, *Ascochyta* blight is perhaps the most important worldwide constraint on chickpea production (Siddique et al., 2012).

This basic listing, although extensive, is an incomplete view of a dynamic crop condition with plants susceptible not only to different types of diseases but also to different species in each disease. *Ascochyta* blight has a range of infective species, including *Ascochyta lentis* on lentil, *A. rabiei* on chickpea and *A. fabae* on faba bean, while *Mycosphaerella pinodes* infects field pea. Crossover between these species is likely and *Mycosphaerella* is advised to be a suitable terminology for the *Ascochyta* blight complex (Gossen et al., 2011).

These distinctions themselves are increasingly incomplete as varieties of virulence continually appear, making protection an increasingly mobile process dependent on rapid monitoring and identification. Each group of pathogens has their own elements of this, with viruses mutating faster than bacteria and fungi, bacteria multiplying rapidly in suitable conditions but less dependent on airborne transmission. Fungi grow more slowly but have extensive sporulation and greater resistance to chemical treatment because of their physiological similarities to plant hosts.

3 Traditional vs. integrated disease management

3.1 Traditional disease management

Developing a disease management system requires both broad protection and, when circumstances arise, specific responses to localised epidemics. Disease management involves an integrated approach ranging from selection of resistant varieties to soil management

© Burleigh Dodds Science Publishing Limited, 2018. All rights reserved.

Table 1 Major diseases of grain legumes

Pathogen	Disease	Crop
Ascochyta fabae	Ascochyta blight	Faba bean
Ascochyta lentis	Ascochyta blight	Lentil
Ascochyta rabei	Ascochyta blight	Chickpea
Aphanomyces euteiches	Common root rot	Pea, faba bean, lentil
Colletotrichum lindemuthianum	Anthracnose	Common bean, chickpea
Erysiphe polygoni	Powdery mildew	Pea
Fusarium solani	Fusarium root rot	Pea, chickpea
Fusarium oxysporum f. sp.	Fusarium wilt	Chickpea, pea, lupin, faba bean, lentil
Macrophomina phaseolina	Dry root rot	Chickpea
Mycosphaerella pinodes	Ascochyta blight	Pea
Peronospora viciae	Downy mildew	Pea
Phoma medicaginis var. pinodella	Foot rot	Pea
Pythium spp.	Seed, seedling and root rot	Pea, chickpea, lentil, common bean, lupin
Rhizoctonia solani	Seedling blight	Lentil, pea, lupin
Sclerontinia sclerotiorum	Stem rot	Lentil, common bean, lupin
Uromyces viciae-fabae	Rust	Pea, faba bean, lentil
Uromyces ciceris-arietini	Rust	Chickpea
Bacteria diseases		
Botrytis cinerea	Botrytis grey mould	Chickpea, lupin
Botrytis fabae, B. cinerea	Chocolate spot	Faba bean
Viral diseases		
Cowpea golden mosaic	Cowpea golden mosaic virus	Cowpea
Sterility mosaic	Pigeonpea sterility mosaic virus	Pigeonpea
Yellow vein mosaic	Yellow vein mosaic	Mung bean

in advance of planting to pesticide application during cultivation. Certain critical disease sources such as seed infection (Chang et al., 2007) are well recognised and can be addressed centrally, while others, such as climate dependency, vary according to location.

For some time pesticides provided unqualified support to the aim of increasing crop yields and allowed extensive industrialisation of production (Clark, 2006). In a parallel to the effectiveness of clinical antibiotics, disease resistance in plant pathogens developed rapidly, leading to a progressive increase in pesticide production – from a negligible global level in 1945 to three million tons per year in 1985 (Tilman et al., 2002). Because of the selection pressure exerted by extensive antibiotic applications, microorganisms can

© Burleigh Dodds Science Publishing Limited, 2018. All rights reserved.

develop resistance to antibiotics within one to three years of their application requiring continual refinement of treatments.

The disease process in plants is a mirror of clinical disease development with progression stages from transmission, exposure, adherence, infection, colonisation and growth, and finally symptoms. Each of these stages has the potential for control and a constant battle between pathogen and host has driven the evolution of increasing virulence and concomitant host response for millennia.

3.2 Integrated disease management

It has become increasingly appreciated that no one solution can address the complexities of multiple disease incidence on crops (Makkouk and Kumari, 2009), even with the development of resistant varieties (Pande et al., 2006; Pande et al., 2005). Pesticide treatment alone is increasingly infeasible.

Integrated disease management (IDM) does not seek to remove pesticides entirely but to optimise their effectiveness and minimise their application (Stoddard et al., 2010; Davidson and Kimber 2007). This may be conducted in a number of ways, some direct and individual, others as part of a complex matrix of actions. IDM is also an essential tool for farming in resource-poor locations where pesticide purchase, advanced testing technology or even mechanisation may be limited. Moreover, legumes are more sensitive to stress factors than other crops because of the complex relationship with *Rhizobium* bacteria (Bordeleau and Prevost, 1994) and require more attention on a continual basis.

Similar approaches to disease prevention have been developed in other areas, for example, hurdle technology in food preservation, where a series of barriers provides challenges to microbial growth or degenerative processes (Sing and Shalini, 2016; Leistner, 2000). For example, in beer stabilisation, the combined effect of low pH, high acidity, alcohol, low oxygen and hop compounds provides very effective protection against the development of pathogens (Vriesekoop et al., 2012). In other foods, elements of these hurdles such as citric acid or salt are added in formulation to assist intrinsic factors.

IDM need not be complex as a single intervention may incorporate a number of effects. An example of a simple initiative to minimising pesticide application is reported from Yunnan Province in China using the co-culturing of rice varieties susceptible and resistant to rice blast disease (*Magnaporthe grisea*). Planting alternative rows of the two varieties resulted in a 94% reduction in disease severity and 89% greater yield of the susceptible variety than when in monoculture (Zhu et al., 2000). This programme was so successful that fungicide applications were stopped after two years. A variety of mechanisms may explain these results which do need further analysis before extrapolation. Barrier exclusion may be a simple option but also incorporate effects on microclimate. Possible transmission of endophytes between plants may be a more complex possibility. Other mechanisms to explain the impact of intercropping plants include the production of volatile organic compounds by trap plants to attract predators carrying pathogens (Stenberg et al., 2015). Such volatiles may also be considered as markers to indicate plant disease and feasibly could be bioengineered into crops as indicator systems (Lucas, 2011). In many cases of IDM, however, more complex management systems are developed and applied, often targeted to specific crops or diseases.

A. rabiei blight is a major crop infection in chickpea and other legume crops attacking leaves, stems and pods. Infection can develop at all stages, particularly during flowering,

© Burleigh Dodds Science Publishing Limited, 2018. All rights reserved.

resulting in up to 100% losses in susceptible varieties. Cereals are a major host of many *Ascochyta* species, but some are found to be pathogenic only on specific legume species (Ahmed and Beniwal, 1991). Control is possible with targeted fungicide treatment (Chongo et al., 2003) but benefits greatly from a multilayer approach to integrate prediction and active prevention. This is increasingly desirable as increasing use of systemic fungicides appears to carry a risk of increased fungicide insensitivity (Gossen et al., 2014).

Dispersal of inoculum by rainsplash from infected field debris (Pedersen et al, 1994) and active foliage infection are major sources of disease spread and greatly aided by persistent wetness followed by dry periods with optimal infection temperatures of 20°C for young seedlings. Spores can persist for up to two years in low humidity conditions below 65% relative humidity (RH), but viability declines if diseased tissue is buried below a 10-cm depth (Kaiser, 1973). Ascospores are, however, more infective than conidiospores under stress conditions, thus making physiology of infection on plant residues an important feature (Trapero-Casas and Kaiser 2007). Pathogen genetics of *A. rabiei* in Canada indicates high levels of diversity, suggesting that variations in virulence may pose challenges to static management practices as more virulent strains of pathogen arise (Chongo et al., 2004). A similar extent of variation was also noted in isolates from the western provinces of Iran (Vafaei et al., 2016).

With specific knowledge of a pathogen physiology and virulence, it is increasingly possible to develop an IDM to both predict disease potential and act on its presence. In the next section, we review the components which are potential elements of an integrated approach to any disease control, using *Ascochyta* as an example.

4 Components of IDM

4.1 Selection of variety

Selection of seed variety has a major impact on disease incidence (Chang et al., 2007), but the mechanism for this is difficult to elucidate without details of pathogen virulence and host resistance (Rubiales and Fondevilla, 2012).

Identification of potential resistant varieties has traditionally depended on identifying resistant plants in the field, but this has difficulties as disease virulence changes (Tivoli et al., 2006). Advanced techniques have been reported to identify potential resistance genes (Ohm et al., 2012; Skiba et al., 2005; Satovic et al., 2003) but more extensive genetic information is required to apply this information into a practical context, particularly as many genes code for partial resistance (Rubiales and Fondevilla, 2012; Jayakumar et al., 2005; Tar'an et al., 2003).

High-throughput phenotyping of plant features is also being developed to help selection of varieties with improved growth and resistance traits. These traits may be biochemical but also physiological such as root architecture (Burridge et al., 2016).

In addition, it is relevant that as disease incidence increases with plant stress (Bostock et al., 2014; Schoeneweiss, 1975), resistant varieties are tested to cover such growing conditions, particularly as, for example, chickpea yield is improved if plants are drought stressed in later growth stages to induce increased root growth (Kashiwagi et al., 2014).

It is further evident that infection of plants with one disease may increase vulnerability of seeds to other diseases (Hepperly, 1979). Developing disease resistance to one pathogen

© Burleigh Dodds Science Publishing Limited, 2018. All rights reserved.

need not necessarily provide protection against another, possibly because of the disease trade-off noted earlier (McGrann et al., 2014).

A further dimension of pathogen control and variety selection is the presence of endophyte microorganisms providing intrinsic protection. Endophyte microorganisms have been noted to be closely related to plant pathogens and may have derived a commensal relationship to the benefit of both partners (De Meyer et al., 2015: Schardl et al., 1997). Endophytes may have many potential modes of action (Dedeja et al., 2012), but specific antagonism to pathogenic microorganisms has been documented for *Pseudomonas* species (O'Sullivan and O'Gara, 1992). Endophytes have also shown action as growth promoters (Zhao et al., 2013). Beneficial inoculation of chickpea with endophyte bacteria has been reported and the potential for including endophytes in addition to *Rhizobia* in legume seeds is a valuable area of development (Maltzahn et al., 2016: Rajendran et al., 2015).

4.2 Seed quality

Infected seeds are a fundamental source of plant pathogenesis (Burgess et al., 1997; Bretag et al., 1995) and, if based in a major production facility, can disseminate disease to a broad area. Locating seed production away from a known infected area may help reduce the potential for infection, but hygienic production facilities are critical to achieve a target of less than 0.3% infection. Management of production is also important as damage to seeds in processing will make seeds vulnerable to infection. Large-seeded legumes are particularly susceptible to damage in handling due to their thin testa which are easily cracked and having embryo parts close to the seed coat. Harvesting seed crops, thus, require mechanical surfaces to be protected and impact to be minimised.

Additional controls for seed hygiene commonly include addition of fungicides and fumigation (Kaiser, 1982) but also heat treatment to reduce pathogen vigour (Ahmed and Beniwal, 1991; Zinnen, 1982). The use of agents to stimulate natural defence activators or elicitors has also been developed and patented (Eulgem et al., 2016; Melander and Rogers, 2016) as has been the application of small-interfering RNAs (siRNAs) (Huang et al., 2016). A caution on the use of fungicides on seeds is to ensure that they are compatible with the addition of *Rhizobium* inoculant.

4.3 Location

Choice of field location for cultivation has broad and narrow options of control. Areas with moist warm climates are more likely to induce *Ascochyta* growth (Casas and Kaiser, 1992), while fields with a history of disease will clearly have a reservoir of infection. A three or four species crop rotation may minimise the latter but will have limited impact if wind dispersal is likely from more distant locations (Salam et al., 2011; Madden, 1997; Peddersen and Morrall, 1995). Ideally, a separation of 500 metres from fields with an infection history is desirable along with bordering fields hosting non-infective crops.

4.4 Field hygiene

Where growing location options are constrained, an active management of crop residues and soil quality can be important factors in limiting infection. Removal or combustion of infected residues will reduce infection sources but is difficult to achieve fully as soil will

© Burleigh Dodds Science Publishing Limited, 2018. All rights reserved.

remain a reservoir. Soil itself is not inert and, as with the current awareness of human intestinal biome, active microorganisms in soil have been shown to have suppressive effects (Camsos et al., 2016; Mazzola, 2002). Management practices to enhance commensal microbes with disease suppression indicate that no-tillage systems had the greatest effect in reducing disease by *Fusarium graminearum* (Campos et al., 2016).

4.5 Biocontrol

Fluorescent *Pseudomonas* species have been suggested as being a facilitator in the suppression of plant diseases by the production of the antibiotic 2,4-diacetylphloroglucinol (Mazzola, 2002), and antibiotics have long been noted as viable disease control agents (Handelsman and Stabb, 1996). *Rhizobium* inoculation has also been noted as reducing the development of disease in legumes (Hemissi et al., 2013). Antibiotic-mediated suppression of pathogens has been elucidated to specific mechanisms such as interfering with zoospore-homing events in *Pythium* infection of pea (Heungens and Parke, 2000).

More intimate measures to control infection include modifying the plant genome to increase resistance. This, for example, can include increasing intrinsic barriers such as root cell walls and phytoalexin production (Lozavaya et al., 2005). One issue in altering intrinsic resistance is the balance within the plant genome between resistance against biotrophic and necrotrophic pathogens and the possibility of disrupting a trade-off between these (Erb et al., 2011). Nevertheless, a number of promising avenues are being explored (Chen et al., 2012), particularly control of hormone production, for example, abscisic acid, defence modulators such as NPR1 (Kim and Delaney, 2002) and transcription factors controlling gene activation (Gurr and Rushton, 2005).

Of course, plant disease is not all microbial, parasites and pests are equally effective in depleting growth and yields. Nematodes are well-known soil parasites of legumes (Davis et al., 2005), and the possibility of identifying mutants with enhanced resistance has been developed using Target-induced Local Lesions in Genomes (TILLING) technologies (Perry et al., 2003).

4.6 Plant density and arrangement

Ascochyta germination and infection is greatly affected by moisture and can easily be enhanced by canopy humidity (Pedersen and Morrall, 1994). Planting density and positioning can thus be important controls to limit disease initiation and spread both from direct infection and via insect vectors (Hema et al., 2014). *Ascochyta* incidence is noted to correlate with plant density (Chang et al., 2007). Density of planting is a well-established factor for crop yield with, for example, between 35 and 70 plants per m^2 recommended for chickpea in Canada (Gan et al., 2003) and 40 plants per m^2 in Queensland, Australia (Beech and Leach, 1989). Because chickpea initially grows slowly compared to other plants, a moderate to high density is advantageous in reducing weed competition. However, impacts on disease spread may conflict with yield targets. For example, wide row spacing has been noted to reduce *Ascochyta* incidence (Chang et al., 2007) but will also reduce yields, making a careful calculation necessary to achieve optimal planting for final yield.

Studies also indicate that the time of sowing can affect disease incidence, with later sowing dates producing less *Ascochyta* disease in field pea in Australia (Bretag et al., 2000), although yields may be reduced, making it a difficult factor to balance without local experience.

© Burleigh Dodds Science Publishing Limited, 2018. All rights reserved.

4.7 Pesticide application

Pesticide use has been curtailed in the face of increased awareness of toxin effects (Eddleston et al., 2002) and in the light of knowledge of how more varied applications may be effective (Madonald and Richardson, 2006). Early fungicides were profoundly broad spectrum and highly toxic but with increasing appearance of pathogen resistance have become more targeted and formulated to allow mixed application at different times and seasons of application (Gossen et al., 2014).

Preventative sprays at early stages of disease development are recommended, but this does require effective crop monitoring and attention to patterns of disease in local areas. Attention to two additional factors, timing and dosing, can help achieve a successful outcome (Banniza et al., 2011).

4.8 Education

A wealth of support services is available to provide information and guidance to farmers to minimise and treat legume diseases, on both local and international bases.

However, past experience and traditional beliefs may be difficult to change. A belief in the power of pesticides or lack of knowledge of disease features may limit IDM (Schreinemachers et al., 2015: Abtew et al., 2016).

4.9 Examples

The above components are all potential elements of an integrated approach to any disease control. They can be applied at local and international levels and are the basis of many collaborative initiatives to predict and manage food security.

Local initiatives often couple with agronomic advice for local areas, for example, Saskatchewan Pulse Growers (http://saskpulse.com/growing/faba-beans/disease-management/). Broader initiatives such as the US Government's Feed the Future programme have a long-term international perspective incorporating research institutes and universities (https://feedthefuture.gov/sites/default/files/resource/files/ftf_factsheet_fsiclegumes_may2015.pdf) and which has a targeted legume programme.

The CGIAR Systemwide Program on Integrated Pest Management (SP-IPM) is a further example of an international collaboration between specialists with broad aims, including anticipating climate change effects and sustainable agriculture (http://www.spipm.cgiar.org/home).

5 Practical developments: modelling, sampling and identification

Looking at future prospects for disease control raises the questions of how IDM may develop and how it may take advantage of advanced technical inputs. As the term is defined, it incorporates sound practical management of existing techniques but can greatly benefit from additional elements of monitoring and prediction, both of which have specific technologies to apply – modelling and genetic analysis.

© Burleigh Dodds Science Publishing Limited, 2018. All rights reserved.

5.1 Modelling

Modelling has both academic relevance and practical application in disease control. Academic studies using experimental data are essential to identify elements in disease development and can include fundamental investigations of pathogen physiology and field exposures.

Pathogen exposure is not essential for model development. Some modelling is based on virtual pathogenesis where a plant's features are adjusted to determine disease dynamics and epidemiology and are a good theoretical basis for field studies (Wilson and Chakrabrty, 1998). Similarly, features of infection such as germination speed can be based on laboratory experimentation (Balkaya, 2004).

In practice, models can allow for direct disease forecasting using field observations of mesoclimatic conditions and daily infection values (Schoeny et al., 2007) and also for broader-scale predictions of continental epidemics (Isard et al., 2007).

Climatic analysis can involve different data sets where inputs can be provided from direct measurement or from regional meteorological information (Palmeri et al., 2006). A study of both of these inputs applied to the same system by Palmeri et al. (2006) indicated similar predictability from both of these sources for fungal disease in turfgrass, suggesting that on-site assessment is not essential, at least for this application. Daily and prior 24-h weather were noted to be the most successful predictors in a study of models for anthracnose disease using data from Australia, Brazil and Colombia when analysed by neural networks (Chakraborty et al., 2004).

Weather data are important factors in model development and, while temperature and humidity are major inputs, other associated data can also feature. Mean RH, number of hours of RH >95%, leaf wetness duration and precipitation during the 48 h prior to sunrise, and minimum and maximum soil temperatures can all be incorporated. The value of timing of leaf wetness is particularly important as these infective periods are strongly correlated to disease incidence by providing free water for germination (Roger et al., 1999; Kaiser, 1973). Although difficult to measure, leaf wetness may be indicated and predicted by high RH.

Direct observation of disease can also be a feature of modelling where threshold levels trigger action to treat a crop or initiate warnings for adjacent cultivation (Adomou et al., 2005; Sikora et al., 2014).

A broader predictability based on accumulating conditions is also possible. Degree-days predictions based on the accumulation of heat units during a growing season are used to predict flowering in crops, and similar approaches can be applied to estimating the potential appearance of disease (Schoeny et al., 2007).

5.2 Sampling

Predication of disease may be based on weather modelling, but on-site sampling allows a more direct and definitive assessment of a location. Visual assessment of crops can indicate the beginnings of an outbreak but does require experienced and skilled observation as well as a rigorous sampling plan to cover the target areas. Too few samples and a localised outbreak may be missed. Too many samples and assessors may be overwhelmed.

In this regard, it is perhaps fortunate that many legume diseases are wind borne and may be detected by air sampling. The target point for air sampling is not exclusively the

© Burleigh Dodds Science Publishing Limited, 2018. All rights reserved.

crop area itself but may be some distance away and most usefully part of a continual monitoring programme (Kennedy and Wakeham, 2008).

Air sampling is typically conducted with mechanised air sampling devices passing a set volume of air over an agar plate, a sticky tape or into tubes which can be processed for molecular analysis (West et al., 2008). Tests may be conducted for fungal spores, bacteria or viruses and with rapid analysis provide a result within 24 h but potentially less (West and Kimber, 2015). Automation can schedule sampling at regular intervals allowing changes in contaminants to be plotted (Kaczmarek et al., 2008).

5.3 Identification

Application of models has considerable potential to predict and manage crop diseases. However, models do depend on accurate knowledge of the presence of disease and this relies on obtaining rapid and correct identification. In addition, being able to detect further features of virulence or strain changes require both advanced technologies and rapid response, ideally as close to field conditions as possible rather than relying on distant, and delayed, laboratory testing.

This latter demand is difficult to achieve with traditional techniques. Classical identification of field crop diseases has relied initially on observation by growers and on evaluations by experienced practitioners. Unfortunately, visual symptoms are frequently obtained late in the disease cycle and in many cases too late to avoid crop losses. They do, nevertheless, form part of a broader system to alert nearby and more distant locations to the hazard as well as help direct action on the crop itself.

Difficulties in crop observation are evident in the need for experienced practitioners to make assessments and, in many cases, for access to technical resources for confirmation. Developments to overcome this include using machine learning and neural networks to provide identification of images forwarded from field observations. Public-access archives of photographs of healthy and diseased plants can be used for assessment by appropriate algorithms to provide direct feedback, for example PlantVillage developed by Hughes and Salathé from Penn State College of Agricultural Sciences in 2012 (https://www.plantvillage.org/). Remote surveillance is an extension to this where airborne drones may provide the input from a distance (Lucas, 2011).

Disease symptoms may overlap with stress symptoms compounding confusion. The acceleration of infection by stress is a further element enhancing disease progression (Schoenweiss, 1975) and making it essential to act on the presence of symptoms.

Confirmation of field identification relies on practitioners having appropriate training and experience, but subjective elements are inevitably involved and may vary between regions and even between countries due to differences in training and access to contemporary literature. In cases of bacterial disease, a series of staining and biochemical tests may provide identification, particularly if these tests involve a means of rapid development of reagents such as those contained in test strip systems or microtitre plate growth monitoring.

Laboratory assessments are particularly problematic for fungal diseases where fungal species can have complex life cycles often involving multiple differentiation of spores and tissues. Microscope observation requires access to these differentiated features as hyphae of different species may be similar and indistinguishable.

© Burleigh Dodds Science Publishing Limited, 2018. All rights reserved.

With suitable expertise, classical analysis may provide essential information for both direct feedback and epidemiological study, and standard techniques remain in use by analysis services (Montana Regional Pulse Crop Diagnostic Laboratory 2016).

Classical identification procedures have further difficulties in that not all species are easily cultured and so cannot provide material for observation. Some pathogens may have very specific nutrient needs, some of which may be unknown, making the pathogen unculturable and impossible to identify without advanced techniques.

Extra demands of increased identification and analysis include the need for speed and also to provide data which may be used in broad epidemiological analysis of a disease outbreak. Advanced techniques are not always rapid, particularly if specialist skills are required. These advanced techniques are increasingly prevalent in both investigative and routine studies of phytopathogens. Many are more suitable for academic investigations into specific disease mechanisms, but some have promise of translating into systems suitable for on-site field or local laboratory use.

An additional application is in food safety assessment where legume products are released for consumption and require confirmation of the absence of contamination. These tests are also relevant to assess imported shipments to ensure that they do not carry disease to novel regions.

As well as providing identification of a disease, methods are also required to provide an assessment of its severity and whether mutations may have produced or activated specific virulence factors (Ohm et al., 2012). Quantification of disease may be conducted on soils and plant material as well as on air and water samples. An integrated plan of monitoring can not just identify the locations and sources of disease but profile its development and anticipate potential effect as well. Techniques are available for assessing all of these and so allow rapid response and effective control.

6 Advanced and rapid analysis techniques

The value of advanced techniques in clinical disease management is well recognised and is a major focus for clinical disease prevention. While the specific benefits of these techniques are clear, it is relevant to assess which of the many different approaches are relevant and suitable for long-term application. Given the pace of technical developments, it is likely that many practitioners and disease managers will have relevant instruments in their hands in the not too distant future.

Advances in the past two decades have taken place with particular focus on molecular biology but also with targets to address speed of processing and automation. The ideal for future applications is to have portable equipment which can process small volumes of samples within the time of a field visit and provide a direct diagnosis with a high level of accuracy.

Because legume diseases are easily encountered by airborne distribution, methods to integrate molecular biology with air sampling are particularly relevant (West and Kimber, 2015: Kaczmarek et al., 2008; West et al., 2008; Fraaije et al., 2005).

Additional targets of analysis are to obtain discrete data such as DNA sequences which can be used for future reference and to allow processing in epidemiological studies.

© Burleigh Dodds Science Publishing Limited, 2018. All rights reserved.

Disease success depends on the presence and activation of virulence factors. Identifying the presence of these, irrespective of identification of a disease agent, is a further benefit, particularly in the light of horizontal gene transfer where pathogenesis may move between species.

6.1 Sequence-based methods

Molecular methods have certainly addressed the targets of obtaining discrete data along with speed and automation, albeit with varying degrees of success. Molecular biology based on DNA, RNA or protein is typically the basis of many contemporary procedures, with the molecular sequence of these being the fingerprint providing relevant information. The cornerstone of nucleic acid methods is the polymerase chain reaction (PCR) catalysing amplification of DNA sections corresponding to specific gene sequences. Short primer sequences of DNA are designed to initiate this amplification and are key to making the process specific for target organisms or genes. As such, the design of primers is a critical factor in the success of PCR assays (Robertson and Walsh-Weller, 1998). For species identification, primer sequences based on the ribosomal genes are typically used.

Identification is, inevitably, a major requirement to allow relevant and specific control measures to be enacted. However, this is not always guaranteed. For successful identification, a database is necessary and the sequence of a novel pathogen or one with limited prior study may not be available for comparison. In addition, strains of a pathogen species may have different virulence factors and identification of these may require more specific targeting of genes rather than the universal 16S ribosomal gene sequences commonly used for identification.

Success in identification also varies according to the type of pathogen. Bacteria and viruses may be more efficiently identified than fungi due to their single chromosome character. This is much more applicable to whole-genome sequencing (WGS) which will provide much more information than a single gene profile.

Fungi, being eukaryotic organisms with multiple chromosomes, are less amenable to WGS and require more complex approaches. It is, perhaps, not simple chance that fungi pose more pathogenic risks to crops than viruses and bacteria do. However, the advantage of their more complex genome provides an ability to differentiate and more extensively integrate with the host. Despite this, whole-genome sequences have been obtained for major legume diseases (Dash et al., 2016), with a genomic data portal accessible at http://legumeinfo.org.

While WGS allows high-level identification, it opens additional avenues of analysis which molecular methods may further develop. In particular, it permits extensive comparison of sequences from different samples and therefore allows progress of a disease to be tracked and its location to be specified. It further allows changes to be identified and, perhaps, correlated to selection stresses present in different environments. However, despite many applications in clinical diseases (Berg et al., 2011), costs are likely to make WGS more suitable for academic and international interest than for local legume disease management.

More applicable methods of monitoring legume diseases centre on amplification of short sequences of DNA or RNA, patterns of amplified DNA, immunological reactions or protein mass spectroscopy. These approaches will be considered in comparison, but there are some general issues which molecular tests must address in order to provide reliable use.

© Burleigh Dodds Science Publishing Limited, 2018. All rights reserved.

Despite the potential of molecular methods and their advantages over classical observation and laboratory study, a number of limitations must be addressed for specific tests to be validated. These centre particularly on sensitivity and specificity. Sensitivity is less of an issue when a sample is obtained from a purified laboratory culture but is critical when field and environmental samples are processed. A small pick of a plate culture will provide enough DNA, RNA or protein for a direct analysis by genotyping, PCR or protein MALDI TOF.

In contrast, an environmental sample may contain only a few microorganisms and moreover be mixed with other species and with potential inhibitors of the analysis method. Soil is a particularly difficult medium to process due to levels of humus-inhibiting PCR polymerase enzymes (Fatima et al., 2014), but plant material can be similarly challenging (Maropolo et al., 2015). Extracting and purifying the target species becomes a limiting factor, particularly when hard cell walls or spores require digestion. Early classical extraction techniques using phenol and alcohol precipitation (Yeates et al., 1997) have been superseded by optimised enzyme-based kits from many commercial suppliers, but low microbial levels may require concentration of samples or bulking.

While sensitivity addresses the limits of detection, specificity is needed to distinguish one species from another but is limited by the uniqueness of the features chosen to be assessed. These may vary from the whole genome as in WGS to short sections of interspecific spacer areas of the ribosomal gene to the pattern of amplified DNA bands produced by primers amplifying random areas of the genome. In all cases for identification, a database is required to match the produced fingerprint to ones with an agreed taxonomic identification.

Databases are required for comparison to previously identified samples deposited as type samples. Some databases such as GenBank at the National Centre for Biotechnology Information (http://blast.ncbi.nlm.nih.gov/blast/Blast.cgi) are extensive, covering the full extent of prokaryotes and eukaryotes for DNA and protein sequences. Some are focused on disease organisms in general, such as the International Society for Human and Animal Mycology (http://its.mycologylab.org). Others may be more targeted to specific genera such as the *Aspergillus* website (http://www.aspergillus.org.uk/) and thus contain more detailed information on strains as well as information on recent developments, methods and pathogenesis.

Sites dedicated to specific diseases such as *Aspergillus* or *Fusarium* (http://www.fusariumdb.org/index.php) provide genomic database libraries with search and compare features to allow such identifications and also to log entries for future comparisons. More specific sites such as the Pathogen Host Interactions (PHI-base, www.phi-base.org) focus on disease genes to assist targeting anti-infective chemistries. The Legume Information System (http://legumeinfo.org/genomes) hosts the genome data of seven legume species at present (2016) and can be a resource of information for resistance genes.

An essential caution of databases is, of course, that the reference and type samples are accurately identified by classical means. In addition, skills are required in interpreting the results of database returns to avoid false positives and false negatives (Foldfeder et al., 2016). Difficulties with accuracy were noted in early use of databases (Clayton et al 1995) and in surveys since (Ashelford et al., 2005; Janda and Abbott, 2007), but more rigorous screening will increase reliability.

© Burleigh Dodds Science Publishing Limited, 2018. All rights reserved.

Sensitivity and specificity in molecular biology analysis depends on careful development of methods. Many species have similar sequences in their genes and identifying which area is most discriminatory requires *in silico* analysis and careful primer design. The internal transcribed spacer (ITS) region of the ribosomal genes in fungi are routinely targeted for primer design and provide suitable sources of variation to identify species (Manter and Vivanco, 2007). Sensitivity is also affected by primer design and amplification method, thus requiring validation to determine minimum copy number for a reliable response. This can make a difference if low numbers of cells are to be detected, such as on air sample filters (West et al., 2008).

Bacteria identification also uses ribosomal genes, typically those of the 16S ribosome. The value of this location was recognised early for clinical use (Clarridge 2004) and has since been applied to most areas of bacterial analysis (Chakravorty et al., 2007). As with fungi, the choice of ribosomal genes has the typical advantage of the presence of multiple copies, thus making the test more sensitive than if a specific genome were chosen. A similar rationale applies to targeting mitochondrial genes. Ribosomal RNA may also be amplified and has advantages of being applicable to environmental samples with multiple organisms. The higher copy numbers of the RNA providing a higher starting concentration of nucleic acid and thus a greater sensitivity (Moreno-Paz and Parro, 2006).

Extensions to the basic PCR procedure provide greater precision (Srinivasa et al., 2012) and it is likely that further developments will be applied to more specific genera as genome information becomes more extensively studied. Other developments in methodology have focused on optimising amplification of microbial DNA in the presence of plant tissue as is required in disease analysis of live samples (Tsoktouridis et al., 2014).

6.2 Genotyping methods

Sequencing is not the only means by which a DNA sample may be assessed. Amplification of target parts of the genome is directed by the choice of primer, but the amplified DNA may be compared directly. For this the DNA may be separated and viewed as band patterns on an agarose or on a polyacrylamide gel. The size of a band is indicative of the target gene and can be taken as indicative of a suspect organism if the primer is totally specific for a gene present only in that organism.

In a less definitive version of standard PCR, the use of more random primers may produce a range of amplified DNA products of different size. This random amplified polymorphic DNA (RAPD) technique uses short primers of 9–11 base pairs to scan whole genomes. When run on a gel, the amplified DNA produce a definitive pattern which may be compared to database sets for identification (Irinyi et al., 2016: Micheli et al., 1994). A major limitation of this approach is that the sample must be pure when isolating DNA as DNA of other organisms, including operators, will confuse the banding.

In more complex analysis, a purified sample of amplified DNA can be digested by nuclease enzymes and the digest profiled on a gel to obtain a fingerprint. Amplified fragment length polymorphism (AFLP) has been widely used to genotype fungi species (Bensch and Akesson, 2005) and also to identify resistance markers in host plants (Ajay et al., 2015).

Identification of individual pathogens is, inevitably, a simplification of the microbial community present at any one time in a plant's life cycle and similarly of the microbiome in the soil and environment. Next-generation sequencing can provide a broader view of these communities, not always at species level but specific enough to provide data for

© Burleigh Dodds Science Publishing Limited, 2018. All rights reserved.

trend analyses. Its use in fingerprinting genomes of pathogens is particularly valuable (Barba et al., 2014) and has been applied to legume viruses (Kehoe et al., 2014).

6.3 Quantitative techniques

Determining the levels of a disease organism can be equally important as detecting its presence. Routine PCR will produce amplified DNA, but some techniques can quantify the amount and relate this to the numbers of copy number and microbial cells initially present.

Real-time PCR (RT PCR) is perhaps the most prevalent method to do this and is extensively applied to many quality assurance processes for food safety as well as clinical analysis. RT PCR follows the same protocols as standard PCR but is able to measure the increase in DNA amplification with time. The amplification system involves the inclusion of a component containing fluorescent probes and quenchers. Because of the action of the TAQ polymerase enzyme, the fluorescent probe is activated during the amplification in proportion to the DNA produced. This will in turn depend on the amount of DNA present at the start of the reaction and so allow a quantification, either in comparison with another product or for absolute measure against a standard curve of purified target DNA.

Applications of RT PCR are increasingly evident in disease analysis both for quantification of a pathogen (Kaczmarek et al., 2009) and for detection of the presence of a mutation in a fungicide target gene (McCartney et al., 2003).

RT PCR also allows a melt curve of the amplified DNA to be calculated. The temperature and profile of the melt curve will depend on the DNA sequence and can be an indicator for identification. It is becoming increasingly common to combine a number of primers into one multiplex RT PCR reaction so that more than one disease may be assessed at the same time, thus reducing costs. In providing results directly, RT PCR allows for a rapid analysis. Coupled with the development of small-scale instruments, the potential of the technique for field use is increasingly close.

While developments of RT PCR lie in the use of increasingly complex reagents, an alternative approach relies on the use of increasingly complex primers. Loop-mediated isothermal amplification (LAMP) is based on conducting PCR assays at a constant temperature and analysing the result in a simple spectrophotometer to detect an increase in optical density at 600nm as the amplified DNA precipitates pyrophosphate salts (Notomi et al., 2000). LAMP uses a mix of four primers targeting six target sequences on the DNA template and acting sequentially to amplify the DNA into complex three-dimensional structures with multiple loops. Detection by simple optical density allows the technique to be used in more basic conditions than standard PCR and so has the potential for extensive application. The technique has been used as a rapid test for a range of bacterial pathogens and more recently for fungi (Dai et al., 2012) including *Asochyta* in chickpeas (Chen et al., 2016) and *Erwinia* species (Temple and Johnson, 2011).

PCR is not the only approach to rapid analysis. Immunochemical tests such as ELISA have been used for decades with increasing effectiveness. As noted earlier, applications to disease monitoring include coupling automatic air sample systems impacting air samples on microtitre strips to collect samples at regular intervals dictated by the sampling control systems. Subsequent processing allows a rapid development of response to indicate amounts of a target antibody (Rogers et al., 2008; Kaczmarek et al., 2009).

© Burleigh Dodds Science Publishing Limited, 2018. All rights reserved.

7 Conclusion

From the range of considerations outlined here, it is evident that legume disease management has many elements and a strong potential to achieve success. This is, however, assuming the availability of strong and effective management and access to skilled practitioners on farms and in supporting laboratories. It requires local engagement and international initiatives with regional applications.

In particular, it requires the ability to act rapidly and effectively and increasingly so with climate change effects. Speed is often on the side of the disease with ability to germinate in hours, penetrate tissue rapidly and multiply extensively. Legume varieties may have partial resistance but require additional hurdles to suppress pathogen growth. It is the task of all practitioners to keep these hurdles in place and develop more when the track of infection changes course.

8 Where to look for further information

Africa Soil Health Consortium. Legume Crop Pests and Diseases. https://africasoilhealth. cabi.org/materials/legumes-crop-pests-and-diseases/

CGIAR Research Program on Grain Legumes. Annual Report 2015. http://library. cgiar.org/bitstream/handle/10947/4458/Grain-Legumes-Annual-Report-2015. pdf?sequence=1

Food and Agriculture Organization of the United Nations. http://agris.fao.org/agris-search/search.do?recordID=ES2003002116

Legume Federation. https://legumefederation.org/

United States Department of Agriculture. Agricultural Research Service. Grain Legume Research Physiology Research Publications. https://www.ars.usda.gov/research/publications/publications-at-this-location/?modecode=20-90-20-00

9 References

Abtew, A., Niassy, S., Affognon, H., Subramanian, S., Kreiter, S., Garzia, G. T. and Martin, T., 2016. Farmers' knowledge and perception of grain legume pests and their management in the Eastern province of Kenya. *Crop Protect.* 87, 90–7.

Adomou, M., Prasad, P. V. V., Boote, K. J. and Detongnon, J., 2005. Disease assessment methods and their use in simulating growth and yield of peanut crops affected by leafspot disease. *Ann. App. Biol.* 146, 469–79.

Ahmed, S. and Beniwal, S. P. S., 1991. Ascochyta blight of lentil and its control in Ethiopia. *Tropical Pest Manage.* 37, 368–73.

Ajay, B. C., Gowda, M. B., Prasad, P. S., Veerakumar, G. N., Babu, P. H., Ganesh, B. N., Venkatesh, S. C., Ganapathy, K. N., Fiyaz, R. A. and Ramya, K. T., 2015. Identification of AFLP markers linked to Fusarium wilt disease in pigeonpea [*Cajanus cajan* (L.) Millsp.]. *Legume Res.* 38, 126–30.

Ashelford, K. E., Chuzhanova, N. A., Fry, J. C., Jones, A. J. and Weightman, A. J., 2005. At least 1 in 20 16S rRNA sequence records currently held in public repositories is estimated to contain substantial anomalies. *Appl. Environ. Microbiol.* 71, 7724–36.

© Burleigh Dodds Science Publishing Limited, 2018. All rights reserved.

Barba, M., Czosnek, H. and Hadidi, A., 2014. Historical perspective, development and applications of next-generation sequencing in plant virology. *Viruses* 6, 106–36.

Balkaja, A., 2004. Modelling the effect of temperature on the germination speed in some legume crops. *J. Agron.* 3, 179–83.

Banniza, S., Armstrong-Cho, C. L., Gan, Y. and Chongo, G., 2011. Evaluation of fungicide efficacy and application frequency for the control of ascochyta blight in chickpea. *Can. J. Plant Pathol.* 33, 135–49.

Beech, D. and Leach, G., 1989. Effect of plant density and row spacing on the yield of chickpea (cv. Tyson) grown on the Darling Downs, south-eastern Queensland. *Aust. J. Expt. Agric.* 29, 241.

Bensch, S. and Akesson, M., 2005. Ten years of AFLP in ecology and evolution: why so few animals? *Mol. Ecol.* 14, 2899–2914.

Berg, J. S., Khoury, M. J. and Evans, J. P., 2011. Deploying whole genome sequencing in clinical practice and public health: Meeting the challenge one bin at a time. *Genet. Med.* 13, 499–504.

Bordeleau, L. M. and Prévost, D., 1994. Nodulation and nitrogen fixation in extreme environments, in Graham, P. H., Sadowsky, M. J. and Vance, C. P. (eds), *Symbiotic Nitrogen Fixation, Developments in Plant and Soil Sciences*. Springer: Netherlands, pp. 115–25.

Bos, L. and Makkouk, K. M., 1994. Insects in relation to virus epidemiology in cool season food legumes, in: Muehlbauer, F. J. and Kaiser, W. J. (eds), *Expanding the Production and Use of Cool Season Food Legumes, Current Plant Science and Biotechnology in Agriculture*. Springer Netherlands, pp. 305–32.

Bostock, R. M., Pye, M. F. and Roubtsova, T. V., 2014. Predisposition in plant disease: Exploiting the nexus in abiotic and biotic stress perception and response. *Annu. Rev. Phytopathol.* 52, 517–49.

Bouràoui, M., Abbes, A., Abdi, N., Hmissi, I. and Sifi, B., 2012, Evaluation of efficient *Rhizobium* isolates as biological control agents of *Orobanche foetida* Poir. Parasitizing *Vicia Faba* L. minor in Tunisia. *Bulg. J. Agric. Sci.* 18, 557–64.

Bretag, T., Price, T. and Keane, P., 1995. Importance of seed-borne inoculum in the etiology of the Ascochyta blight complex of field peas (*Pisum sativum* L.) grown in Victoria. *Aust. J. Exp. Agric.* 35, 525.

Bretag, T. W., Keane, P. J. and Price, T. V., 2000. Effect of sowing date on the severity of ascochyta blight in field peas (*Pisum sativuum* L) grown in the Wimmera region of Victoria. *Aust. J. Exp. Agric.* 40, 1113–19.

Brown, J. K. M. and Rant, J. C., 2013. Fitness costs and trade-offs of disease resistance and their consequences for breeding arable crops. *Plant Pathol.* 62, 83–95.

Burgess, D. R., Bretag, T. and Keane, P. J., 1997. Seed-to-seedling transmission of *Botrytis cinerea* in chickpea and disinfestation of seed with moist heat. *Aust. J. Exp. Agric.* 37, 223.

Burridge, J., Jochua, C. N., Bucksch, A. and Lynch, J. P., 2016. Legume shovelomics: High – Throughput phenotyping of common bean (*Phaseolus vulgaris* L.) and cowpea (*Vigna unguiculata* subsp, unguiculata) root architecture in the field. *Field Crops Res.* 192, 21–32.

Campos, S. B., Lisboa, B. B., Camargo, F. A. O., Bayer, C., Sczyrba, A., Dirksen, P., Albersmeier, A., Kalinowski, J., Beneduzi, A., Costa, P. B., Passaglia, L. M. P., Vargas, L. K. and Wendisch, V. F., 2016. Soil suppressiveness and its relations with the microbial community in a Brazilian subtropical agroecosystem under different management systems. *Soil Biol. Biochem.* 96, 191–7.

Casas, A. T., 1992. Development of *Didymella rabiei*, the Telomorph of *Ascochyta rabiei* on Chickpea straw. *Phytopathology* 82, 1261–6.

Chakraborty, S., Ghosh, R., Ghosh, M., Fernandes, C. D., Charchar, M. J. and Kelemu, S., 2004. Weather-based prediction of anthracnose severity using artificial neural network models. *Plant Pathol.* 53, 375–86.

Chakravorty, S., Helb, D., Burday, M., Connell, N. and Alland, D., 2007. A detailed analysis of 16S ribosomal RNA gene segments for the diagnosis of pathogenic bacteria. *J. Microbiol. Methods* 69, 330–9.

© Burleigh Dodds Science Publishing Limited, 2018. All rights reserved.

Chang, K. F., Ahmed, H. U., Hwang, S. F., Gossen, B. D., Howard, R. J., Warkentin, T. D., Strelkov, S. E. and Blade, S. F., 2007. Impact of cultivar, row spacing and seeding rate on ascochyta blight severity and yield of chickpea. *Can. J. Plant Sci.* 87, 395–403.

Chen, X., Ma, L., Qiang, S. and Ma, D., 2016. Development of a loop-mediated isothermal amplification method for the rapid diagnosis of *Ascochyta rabiei* L. in chickpeas. *Sci. Rep.* 6, 25688.

Chen, Y.-J., Lyngkjær, M. F. and Collinge, D. B., 2012. Future prospects for genetically engineering disease-resistant plants, in: Sessa, G. (ed.), *Molecular Plant Immunity*. Wiley-Blackwell, New York, pp. 251–75.

Chongo, G., Buchwaldt, L., Gossen, B. D., Lafond, G. P., May, W. E., Johnson, E. N. and Hogg, T., 2003. Foliar fungicides to manage ascochyta blight [*Ascochyta rabiei*] of chickpea in Canada. *Can. J. Plant Pathol.* 25, 135–42.

Chongo, G., Gossen, B. D., Buchwaldt, L., Adhikari, T. and Rimmer, S. R., 2004. Genetic diversity of *Ascochyta rabiei* in Canada. *Plant Dis.* 88, 4–10.

Clark, W. S., 2006. Fungicide resistance: are we winning the battle but losing the war. *Aspects Appl. Biol.* 78, 119–26.

Clarridge, J. E., 2004. Impact of 16S rRNA gene sequence analysis for identification of Bacteria on clinical microbiology and infectious diseases. *Clin. Microbiol. Rev.* 17, 840–62.

Clayton, R. A., Sutton, G., Hinkle, P. S., Bult, C. and Fields, C., 1995. Intraspecific variation in small-subunit rRNA sequences in GenBank: why single sequences may not adequately represent prokaryotic taxa. *Int. J. Syst. Bacteriol.* 45, 595–9.

Dai, T.-T., Lu, C.-C., Lu, J., Dong, S., Ye, W., Wang, Y. and Zheng, X., 2012. Development of a loop-mediated isothermal amplification assay for detection of *Phytophthora sojae*. *FEMS Microbiol. Lett.* 334, 27–34.

Dash, S., Campbell, J. D., Cannon, E. K. S., Cleary, A. M., Huang, W., Kalberer, S. R., Karingula, V., Rice, A. G., Singh, J., Umale, P. E., Weeks, N. T., Wilkey, A. P., Farmer, A. D. and Cannon, S. B., 2016. Legume information system (LegumeInfo.org): a key component of a set of federated data resources for the legume family. *Nucleic Acids Res.* 44, D1181–8.

Davidson, J. A. and Kimber, R. B. E., 2007. Integrated disease management of ascochyta blight in pulse crops. *Eur. J. Plant Pathol.* 119, 99–110.

Davis, E. L. and Mitchum, M. G., 2005. Nematodes. Sophisticated parasites of legumes. *Plant Physiol.* 137, 1182–8.

De Meyer, S. E., De Beuf, K., Vekeman, B. and Willems, A., 2015. A large diversity of non-rhizobial endophytes found in legume root nodules in Flanders (Belgium). *Soil Biol. Biochem.* 83, 1–11.

Denton, M. D., Pearce, D. J., Ballard, R. A., Hannah, M. C., Mutch, L. A., Norng, S. and Slattery, J. F., 2009. A multi-site field evaluation of granular inoculants for legume nodulation. *Soil Biol. Biochem.* 41, 2508–16.

Dudeja, S. S., 2016. Beneficial effects and molecular diversity of endophytic Bacteria in legume and Nonlegumes, in: Singh, D. P., Singh, H. B. and Prabha, R. (eds), *Microbial Inoculants in Sustainable Agricultural Productivity*. Springer India, New Delhi, pp. 245–56.

Dudeja, S. S., Giri, R., Saini, R., Suneja-Madan, P. and Kothe, E., 2012. Interaction of endophytic microbes with legumes. *J. Basic Microbiol.* 52, 248–60.

Eddleston, M., Karalliedde, L., Buckley, N., Fernando, R., Hutchinson, G., Isbister, G., Konradsen, F., Murray, D., Piola, J. C., Senanayake, N., Sheriff, R., Singh, S., Siwach, S. B. and Smit, L., 2002. Pesticide poisoning in the developing world – a minimum pesticides list. *Lancet* 360, 1163–7.

Erb, M., Balmer, D., De Lange, E. S., Von Merey, G., Planchamp, C., Robert, C. a. M., Röder, G., Sobhy, I., Zwahlen, C., Mauch-Mani, B. and Turlings, T. C. J., 2011. Synergies and trade-offs between insect and pathogen resistance in maize leaves and roots. *Plant, Cell Environ.* 34, 1088–1103.

Eulgem, T., Rodriguez-Salus, M. and Knoth, C. M., 2016. United States Patent: 9420790 – Molecules that induce disease resistance and improve growth in plants. 9420790.

Fatima, F., Pathak, N. and Rastogi Verma, S., 2014. An improved method for soil DNA extraction to study the microbial assortment within Rhizospheric region. *Mol. Biol. Int.* 2014.

© Burleigh Dodds Science Publishing Limited, 2018. All rights reserved.

Fondong, V. N., Thresh, J. M. and Zok, S., 2002. Spatial and temporal spread of cassava mosaic virus disease in cassava grown alone and when intercropped with maize and/or cowpea. *J. Phytopathol.* 150, 365–74.

Fraaije, B. A., Cools, H. J., Fountaine, J., Lovell, D. J., Motteram, J., West, J. S. and Lucas, J. A., 2005. Role of Ascospores in further spread of QoI-resistant cytochrome b alleles (G143A) in field populations of Mycosphaerella graminicola. *Phytopathology* 95, 933–41.

Gan, Y. T., Miller, P. R., McConkey, B. G., Zentner, R. P., Liu, P. H. and McDonald, C. L., 2003. Optimum plant population density for chickpea and dry pea in a semiarid environment. *Can. J. Plant Sci.* 83, 1–9.

Ghini, R., Hamada, E. and Bettiol, W., 2008. Climate change and plant diseases. *Sci. Agricola* 65, 98–107.

Goldfeder, R. L., Priest, J. R., Zook, J. M., Grove, M. E., Waggott, D., Wheeler, M. T., Salit, M., Ashley, E. A., 2016. Medical implications of technical accuracy in genome sequencing. *Genome Med* 8, 24.

Gossen, B. D., Hwang, S. F., Conner, R. L. and Chang, K. F., 2011, Managing the ascochyta blight complex on field pea in western Canada. *Prairie Soils Crops J.* 4, 135–41.

Gossen, B. D., Carisse, O., Kawchuk, L. M., Heyden, H. V. D. and McDonald, M. R., 2014. Recent changes in fungicide use and the fungicide insensitivity of plant pathogens in Canada. *Can. J. Plant Pathol.* 36, 327–40.

Gurr, S. J. and Rushton, P. J., 2005. Engineering plants with increased disease resistance: how are we going to express it? *Trends Biotechnol.* 23, 283–90.

Handelsman, J. and Stabb, E. V., 1996. Biocontrol of Soilborne plant pathogens. *Plant Cell* 8, 1855–69.

Hema, M., Sreenivasulu, P., Patil, B. L., Kumar, P. L. and Reddy, D. V. R., 2014. Tropical food legumes: virus diseases of economic importance and their control. *Adv. Virus Res.* 90, 431–505.

Hemissi, I., Mabrouk, Y., Abdi, N., Bouraoui, M., Saidi, M. and Sifi, B., 2013. Growth promotion and protection against *Orobanche foetida* of chickpea (*Cicer aerietinum*) by two *Rhizobium* strains under greenhouse conditions. *Afr. J. Biotechnol.* 12.

Hepperly, P. R., 1979. Predisposition to Seed Infection by *Phomopsis sojae* in Soybean plants infected by Soybean Mosaic virus. *Phytopathology* 69, 846.

Heungens, K. and Parke, J. L., 2000. Zoospore homing and infection events: effects of the biocontrol Bacterium *Burkholderia cepacia* AMMDR1 on two Oomycete pathogens of pea (*Pisum sativum* L.). *Appl. Environ. Microbiol.* 66, 5192–5200.

Huang, X., McNeill, T. and Schweiner, M., 2016. United States Patent: 9434942 – Small interfering RNAS with target-specific seed sequences. 9434942.

Irinyi, L., Lackner, M., de Hoog, G. S. and Meyer, W., 2016. DNA barcoding of fungi causing infections in humans and animals. *Fungal Biol.* 120, 125–36.

Isard, S. A., Russo, J. M. and Ariatti, A., 2007. The Integrated Aerobiology Modeling System applied to the spread of soybean rust into the Ohio River valley during September 2006. *Aerobiologia* 23, 271–82.

Janda, J. M. and Abbott, S. L., 2007. 16S rRNA gene sequencing for Bacterial identification in the diagnostic laboratory: pluses, perils, and pitfalls. *J. Clin Microbiol.* 45, 2761–4.

Jayakumar, P., Gan, Y. T., Gossen, B. D., Warkentin, T. D. and Banniza, S., 2005. Ascochyta blight of chickpea: infection and host resistance mechanisms. *Can. J. Plant Pathol.* 27, 499–509.

Kaczmarek, J., Fitt, B. F., Jedryczka, M. and Latunde-Dada, A. O., 2008. Detection by real-time PCR and quantification of *Leptospaeria maculans* and *L. biglobosa* in air samples from north Poland. *Aspects Appl. Biol.* 89, 71–6.

Kaiser, W. J., 1983. Etiology and control of seed decay and preemergence damping-off of chickpea by *Pythium ultimum. Plant Dis.* 67, 77.

Kaiser, W. J., 1973. Factors affecting growth, sporulation, pathogenicity, and survival of *Ascochyta rabiei. Mycologia* 65, 444–57.

Kashiwagi, J., Krishnamurthy, L., Crouch, J. H. and Serraj, R., 2006. Variability of root length density and its contributions to seed yield in chickpea (*Cicer arietinum* L.) under terminal drought stress. *Field Crops Res.* 95, 171–81.

© Burleigh Dodds Science Publishing Limited, 2018. All rights reserved.

Kehoe, M. A., Coutts, B. A., Buirchell, B. J. and Jones, R. A. C., 2014. Plant virology and next generation sequencing: experiences with a Potyvirus. *PLoS ONE* 9, e104580.

Kennedy, R. and Wakeham, A. J., 2008. Development and use of detection systems for the sporangia of *Peronospora destructor*. *Aspects Appl. Biol.* 89, 55–6.

Kim, H. S. and Delaney, T. P., 2002. Over-expression of TGA5, which encodes a bZIP transcription factor that interacts with NIM1/NPR1, confers SAR-independent resistance in *Arabidopsis thaliana* to *Peronospora parasitica*. *Plant J.* 32, 151–63.

Koenning, S. R. and Wrather, J. A., 2010. Suppression of Soybean Yield Potential in the Continental United States by Plant Diseases from 2006 to 2009. *Plant Health Progress*. http://www.plantmanagementnetwork. org/sub/php/research/2010/yield/yield.pdf doi:10.1094/PHP-2010-1122-01-RS

Leistner, L., 2000. Basic aspects of food preservation by hurdle technology. *Int. J. Food Microbiol.* 55, 181–6.

Lozovaya, V., Lygin, A., Zernova, O. and Widholm, J., 2005. Genetic engineering of plant root disease resistance by modification of the phenylpropanoid pathway. *Plant Biosyst.* 139, 20–3.

Lucas, J. A., 2011. Advances in plant disease and pest management. *J. Agric. Sci.* 149, 91–114.

Macdonald, O. and Richardson, D., 2006, Regulatory challenges from fungicide resistance. *Aspects Appl. Biol.* 78, 105–12.

Madden, L. V., 1997. Effects of rain on splash dispersal of fungal pathogens. *Can. J. Plant Pathol.* 19, 225–30.

Makkouk, K. M. and Kumari, S. G., 2009. Epidemiology and integrated management of persistently transmitted aphid-borne viruses of legume and cereal crops in West Asia and North Africa. Virus Research, Plant Virus Epidemiology: Controlling epidemics of emerging and established plant viruses – the way forward 141, 209–18.

Manter, D. K. and Vivanco, J. M., 2007. Use of the ITS primers, ITS1F and ITS4, to characterize fungal abundance and diversity in mixed-template samples by qPCR and length heterogeneity analysis. *J. Microbiol. Methods* 71, 7–14.

Maropola, M. K. A., Ramond, J.-B. and Trindade, M., 2015. Impact of metagenomic DNA extraction procedures on the identifiable endophytic bacterial diversity in *Sorghum bicolor* (L. Moench). *J. Microbiol. Methods* 112, 104–17.

Mazzola, M., 2002. Mechanisms of natural soil suppressiveness to soilborne diseases. *Antonie Van Leeuwenhoek* 81, 557–64.

McGrann, G. R. D., Stavrinides, A., Russell, J., Corbitt, M. M., Booth, A., Chartrain, L., Thomas, W. T. B. and Brown, J. K. M., 2014. A trade off between mlo resistance to powdery mildew and increased susceptibility of barley to a newly important disease, Ramularia leaf spot. *J. Exp. Bot.* 65, 1025–37.

Melander, C. and Rogers, S. A., 2016. United States Patent: 9439436 – Use of aryl carbamates in agriculture and other plant-related areas. 9439436.

Micheli, M. R., Bova, R., Pascale, E. and D'Ambrosio, E., 1994. Reproducible DNA fingerprinting with the random amplified polymorphic DNA (RAPD) method. *Nucleic Acids Res.* 22, 1921–2.

Moreno-Paz, M. and Parro, V., 2006. Amplification of low quantity bacterial RNA for microarray studies: time-course analysis of *Leptospirillum ferrooxidans* under nitrogen-fixing conditions. *Environ. Microbiol.* 8, 1064–73.

Murithi, H. M., Beed, F., Tukamuhabwa, P., Thomma, B. P. H. J. and Joosten, M. H. A. J., 2016. Soybean production in eastern and southern Africa and threat of yield loss due to soybean rust caused by *Phakopsora pachyrhizi*. *Plant Pathol.* 65, 176–88.

Narayanasamy, P., 2008. *Molecular Biology in Plant Pathogenesis and Disease Management: Microbial Plant Pathogens*. Springer Science & Business Media, Berlin.

Notomi, T., Okayama, H., Masubuchi, H., Yonekawa, T., Watanabe, K., Amino, N. and Hase, T., 2000. Loop-mediated isothermal amplification of DNA. *Nucleic Acids Res.* 28, E63.

Ohm, R. A., Feau, N., Henrissat, B., Schoch, C. L., Horwitz, B. A., Barry, K. W., Condon, B. J., Copeland, A. C., Dhillon, B., Glaser, F., Hesse, C. N., Kosti, I., LaButti, K., Lindquist, E. A.,

© Burleigh Dodds Science Publishing Limited, 2018. All rights reserved.

Lucas, S., Salamov, A. A., Bradshaw, R. E., Ciuffetti, L., Hamelin, R. C., Kema, G. H. J., Lawrence, C., Scott, J. A., Spatafora, J. W., Turgeon, B. G., de Wit, P. J. G. M., Zhong, S., Goodwin, S. B. and Grigoriev, I. V., 2012. Diverse lifestyles and strategies of plant pathogenesis encoded in the genomes of eighteen *Dothideomycetes* fungi. *PLoS Pathog.* 8, e1003037.

O'Sullivan, D. J. and O'Gara, F., 1992. Traits of fluorescent *Pseudomonas* spp. involved in suppression of plant root pathogens. *Microbiol. Rev.* 56, 662–76.

Palmieri, R., Tredway, L., Niyogi, D. and Lackmann, G. M., 2006. Development and evaluation of a forecasting system for fungal disease in turfgrass. *Met. Apps.* 13, 405–16.

Pande, S., Galloway, J., Gaur, P. M., Siddique, K. H. M., Tripathi, H. S., Taylor, P., MacLeod, M. W. J., Basandrai, A. K., Bakr, A., Joshi, S., Kishore, G. K., Isenegger, D. A., Rao, J. N. and Sharma, M., 2006a. Botrytis grey mould of chickpea: a review of biology, epidemiology, and disease management. *Aust. J. Agric. Res.* 57, 1137–50.

Pande, S., Siddique, K. H. M., Kishore, G. K., Bayaa, B., Gaur, P. M., Gowda, C. L. L., Bretag, T. W. and Crouch, J. H., 2005. Ascochyta blight of chickpea (*Cicer arietinum* L.): a review of biology, pathogenicity, and disease management. *Aust. J. Agric. Res.* 56, 317–32.

Pappa, V. A., Rees, R. M., Walker, R. L., Baddeley, J. A. and Watson, C. A., 2011. Intercropping of legumes and cereals: effect on yield and N uptake in a three year low input crop rotation.

Pedersen, E. A., Morrall, R. a. A., McCartney, H. A. and Fitt, B. D. L., 1994. Dispersal of conidia of *Ascochyta fabae* f. sp. *lentis* from infected lentil plants by simulated wind and rain. *Plant Pathol.* 43, 50–5.

Pedersen, E. A. and Morrall, R. A. A., 1995. Effect of wind speed and direction on horizontal spread of ascochyta blight of lentil. *Can. J. Plant Pathol.* 17, 223–32.

Perry, J. A., Wang, T. L., Welham, T. J., Gardner, S., Pike, J. M., Yoshida, S. and Parniske, M., 2003. A TILLING reverse genetics tool and a web-accessible collection of mutants of the legume *Lotus japonicus*. *Plant Physiol.* 131, 866–71.

Rajendran, G., Sing, F., Desai, A. J. and Archana, G., 2008. Enhanced growth and nodulation of pigeon pea by co-inoculation of Bacillus strains with *Rhizobium* spp. *Bioresource Technology*, exploring horizons in biotechnology. *Global Venture* 99, 4544–50.

Robertson, J. M. and Walsh-Weller, J., 1998. An introduction to PCR primer design and optimization of amplification reactions. *Methods Mol. Biol.* 98, 121–54.

Roger, C., Tivoli, B. and Huber, L., 1999. Effects of temperature and moisture on disease and fruit body development of Mycosphaerella pinodes on pea (*Pisum sativum*). *Plant Pathol.* 48, 1–9.

Rubiales, D. and Fondevilla, S., 2012. Future prospects for Ascochyta blight resistance breeding in cool season food legumes. *Front Plant Sci.* 3, 1–5.

Salam, M. U., Galloway, J., Diggle, A. J., MacLeod, W. J. and Maling, T., 2011. Predicting regional-scale spread of ascospores of *Didymella pinodes* causing ascochyta blight disease on field pea. *Aust. Plant Pathol.* 40, 640–7.

Satovic, Z., Avila, C. M., Cruz-Izquierdo, S., Díaz-Ruíz, R., García-Ruíz, G. M., Palomino, C., Gutiérrez, N., Vitale, S., Ocaña-Moral, S., Gutiérrez, M. V., Cubero, J. I. and Torres, A. M., 2013. A reference consensus genetic map for molecular markers and economically important traits in faba bean (*Vicia faba* L.). *BMC Genomics* 14, 932.

Schardl, C. L., Leuchtmann, A., Chung, K.-R., Penny, D. and Siegel, M. R., 1997. Coevolution by common descent of fungal symbionts (Epichloe spp.) and grass hosts. *Mol. Biol. Evol.* 14, 133–43.

Schoeneweiss, D. F., 1975. Predisposition, stress, and plant disease. *Ann. Rev. Phytopathol.* 13, 193–211.

Schoeny, A., Jumel, S., Rouault, F., May, C. L. and Tivoli, B., 2007. Assessment of airborne primary inoculum availability and modelling of disease onset of ascochyta blight in field peas. *Eur. J. Plant Pathol.* 119, 87–97.

Schreinemachers, P., Balasubramaniam, S., Boopathi, N. M., Ha, C. V., Kenyon, L., Praneetvatakul, S., Sirijinda, A., Le, N. T., Srinivasan, R. and Wu, M.-H., 2015. Farmers' perceptions and management

© Burleigh Dodds Science Publishing Limited, 2018. All rights reserved.

of plant viruses in vegetables and legumes in tropical and subtropical Asia. *Crop Prot.* 75, 115–23.

Siddique, K. H. M., Johansen, C., Turner, N. C., Jeuffroy, M.-H., Hashem, A., Sakar, D., Gan, Y. and Alghamdi, S. S., 2012. Innovations in agronomy for food legumes. A review. *Agron. Sustain. Dev.* 32, 45–64.

Sikora, E. J., Allen, T. W., Wise, K. A., Bergstrom, G., Bradley, C. A., Bond, J., Brown-Rytlewski, D., Chilvers, M., Damicone, J., DeWolf, E., Dorrance, A., Dufault, N., Esker, P., Faske, T. R., Giesler, L., Goldberg, N., Golod, J., Gómez, I. R. G., Grau, C., Grybauskas, A., Franc, G., Hammerschmidt, R., Hartman, G. L., Henn, R. A., Hershman, D., Hollier, C., Isakeit, T., Isard, S., Jacobsen, B., Jardine, D., Kemerait, R., Koenning, S., Langham, M., Malvick, D., Markell, S., Marois, J. J., Monfort, S., Mueller, D., Mueller, J., Mulrooney, R., Newman, M., Osborne, L., Padgett, G. B., Ruden, B. E., Rupe, J., Schneider, R., Schwartz, H., Shaner, G., Singh, S., Stromberg, E., Sweets, L., Tenuta, A., Vaiciunas, S., Yang, X. B., Young-Kelly, H. and Zidek, J., 2014. A coordinated effort to manage Soybean rust in North America: a success story in Soybean disease monitoring. *Plant Dis.* 98, 864–75.

Singh, S. and Shalini, R., 2016. Effect of hurdle technology in food preservation: a review. *Crit. Rev. Food Sci. Nutr.* 56, 641–9.

Skiba, B., Ford, R. and Pang, E. C. K., 2005. Construction of a cDNA library of Lathyrus sativus inoculated with *Mycosphaerella pinodes* and the expression of potential defence-related expressed sequence tags (ESTs). *Physiol. Mol. Plant Pathol.* 66, 55–67.

Srinivasa, C., Sharanaiah, U. and Shivamallu, C., 2012. Molecular detection of plant pathogenic bacteria using polymerase chain reaction single-strand conformation polymorphism. *Acta Biochim. Biophys. Sin.* (Shanghai) 44, 217–23.

Stenberg, J. A., Heil, M., Åhman, I. and Björkman, C., 2015. Optimizing crops for biocontrol of pests and disease. *Trends Plant Sci.* 20, 698–712.

Stoddard, F. L., Nicholas, A. H., Rubiales, D., Thomas, J. and Villegas-Fernández, A. M., 2010. Integrated pest management in faba bean. *Field Crops Res. Faba Beans Sustain. Agric.* 115, 308–18.

Tar'an, B., Warkentin, T., Somers, D. J., Miranda, D., Vandenberg, A., Blade, S., Woods, S., Bing, D., Xue, A., DeKoeyer, D. and Penner, G., 2003. Quantitative trait loci for lodging resistance, plant height and partial resistance to mycosphaerella blight in field pea (*Pisum sativum* L.). *Theor. Appl. Genet.* 107, 1482–91.

Temple, T. N. and Johnson, K. B., 2010. Evaluation of loop-mediated isothermal amplification for rapid detection of *Erwinia amylovora* on pear and apple fruit flowers. *Plant Dis.* 95, 423–30.

Tilman, D., Cassman, K. G., Matson, P. A., Naylor, R. and Polasky, S., 2002. Agricultural sustainability and intensive production practices. *Nature* 418, 671–7.

Tivoli, B., Baranger, A., Avila, C. M., Banniza, S., Barbetti, M., Chen, W., Davidson, J., Lindeck, K., Kharrat, M., Rubiales, D., Sadiki, M., Sillero, J. C., Sweetingham, M. and Muehlbauer, F. J., 2006. Screening techniques and sources of resistance to foliar diseases caused by major necrotrophic fungi in grain legumes. *Euphytica* 147, 223–53.

Trapero-Casas, A. and Kaiser, W. J., 2007. Differences between Ascospores and Conidia of *Didymella rabiei* in spore germination and infection of Chickpea. *Phytopathology* 97, 1600–7.

Tsoktouridis, G., Tsiamis, G., Koutinas, N. and Mantell, S., 2014. Molecular detection of bacteria in plant tissues, using universal 16S ribosomal DNA degenerated primers. *Biotechnol Biotechnol Equip* 28, 583–91.

Uzokwe, V. N. E., Mlay, D. P., Masunga, H. R., Kanju, E., Odeh, I. O. A. and Onyeka, J., 2016. Combating viral mosaic disease of cassava in the Lake Zone of Tanzania by intercropping with legumes. *Crop Prot.* 84, 69–80.

Vafaei, S. H., Rezaee, S., Moghadam, A. A. and Zamanizadeh, H. R., 2015. Virulence diversity of *Ascochyta rabiei* the causal agent of Ascochyta blight of chickpea in the western provinces of Iran. *Arch. Phytopathol. Plant Prot.* 48, 921–30.

© Burleigh Dodds Science Publishing Limited, 2018. All rights reserved.

von Maltzahn, G., Flavell, R. B., Toledo, G. V., Leff, J. W., Samayoa, P., Marquez, L. M., Johnston, D. M., Djonovic, S., Millet, Y. A., Sadowski, C., Lyford, J., Ambrose, K. V. and Zhang, X., 2016. United States Patent: 9408394 – Endophytes, associated compositions, and methods of use thereof. 9408394.

Vriesekoop, F., Krahl, M., Hucker, B. and Menz, G., 2012. 125th Anniversary review: Bacteria in brewing: the good, the bad and the ugly. *J. Inst. Brew.* 118, 335–45. doi:10.1002/jib.49

West, J. and Kimber, R., 2015. Innovations in air sampling to detect plant pathogens. *Ann. Appl. Biol.* 166, 4–17.

West, J. S., Atkins, S. D., Emberlin, J. and Fitt, B. D. L., 2008a. PCR to predict risk of airborne disease. *Trends Microbiol.* 16, 380–7.

Wilson, P. A. and Chakrabrty, S., 1998. The virtual plant: a new tool for the study and management of plant diseases. *Crop Prot.* 17, 231–9.

Wrather, A. and Koenning, S., 2009. Effects of diseases on Soybean yields in the United States 1996 to 2007. *Plant Health Progress.*

Yeates, C., Gillings, M. r., Davison, A. d., Altavilla, N. and Veal, D. a., 1997. PCR amplification of crude microbial DNA extracted from soil. *Lett. Appl. Microbiol.* 25, 303–7.

Zhao, L. F., Xu, Y. J., Ma, Z. Q., Deng, Z. S., Shan, C. J. and Wei, G. H., 2013. Colonization and plant growth promoting characterization of endophytic *Pseudomonas chlororaphis* strain Zong1 isolated from *Sophora alopecuroides* root nodules. *Braz. J. Microbiol.* 44, 623–31.

Zhu, Y., Chen, H., Fan, J., Wang, Y., Li, Y., Chen, J., Fan, J., Yang, S., Hu, L., Leung, H., Mew, T. W., Teng, P. S., Wang, Z. and Mundt, C. C., 2000. Genetic diversity and disease control in rice. *Nature* 406, 718–22.

Zinnen, T. M., 1982. Thermotherapy of Soybean seeds to control Seedborne Fungi. *Phytopathology* 72, 831.

© Burleigh Dodds Science Publishing Limited, 2018. All rights reserved.

CPSIA information can be obtained
at www.ICGtesting.com
Printed in the USA
JSHW030340170821
17892JS00007B/315